Urban Safety and Peacebuilding

This volume draws together original research related to conceptual and practical advances at the interface of urban safety and peacebuilding.

The book reflects the advances in urban safety and peacebuilding to help address the rapidly increasing risk of conflict and insecurity in cities. Specifically, it draws on contributions to the Technical Working Group on the Confluence of Urban Safety and Peacebuilding Practice, an informal expert network co-facilitated by the United Nations Office at Geneva, UN-Habitat's Safer Cities Programme, and the Geneva Peacebuilding Platform. A focus on 'sustaining peace' serves as a framework for situating new policy responses against conflict, violence, and exclusion in the city, and for promoting a conversation across disciplinary and specialist silos. The volume thereby broadens the optic of peacebuilding practice beyond interstate and intrastate armed conflicts – and especially their aftermath – and reconnects it to the community-level origins of building peace. The analysis and practice presented here will remind those willing to work towards peaceful and inclusive cities that there are tried and tested approaches available, and a host of experts and practitioners ready to accompany those prepared to lead in their respective contexts.

This book will be of interest to students and researchers in the fields of peacebuilding, urban studies, security studies, and international relations.

Achim Wennmann is a senior researcher at the Centre on Conflict, Development and Peacebuilding of the Graduate Institute in Geneva, Switzerland, and Executive Coordinator of the Geneva Peacebuilding Platform.

Oliver Jütersonke is Head of Research at the Centre on Conflict, Development and Peacebuilding of the Graduate Institute in Geneva, Switzerland.

Studies in Conflict, Development and Peacebuilding

Series Editors: Keith Krause, Oliver Jütersonke and Riccardo Bocco

Centre on Conflict, Development and Peacebuilding (CCDP), Graduate Institute, Switzerland

This series publishes innovative research into the connections between insecurity and underdevelopment in fragile states, and into situations of violence and insecurity more generally. It adopts a multidisciplinary approach to the study of a variety of issues, including the changing nature of contemporary armed violence (*conflict*), efforts to foster the conditions that prevent the outbreak or recurrence of such violence (*development*), and strategies to promote peaceful relations on the communal, societal and international levels (*peacebuilding*).

Peacebuilding and Ex-Combatants
Political Reintegration in Liberia
Johanna Söderström

Local Ownership in International Peacebuilding
Key Theoretical and Practical Issues
Edited by Sung Yong Lee and Alpaslan Özerdem

Institutional Reforms and Peacebuilding
Change, Path-dependency and Societal Divisions in Post-war Communities
Edited by Nadine Ansorg and Sabine Kurtenbach

Peacebuilding and Spatial Transformation
Peace, Space and Place
Annika Björkdahl and Stefanie Kappler

Exploring Peace Formation
Security and Justice in Post-Colonial States
Edited by Kwesi Aning, M. Anne Brown, Volker Boege, and Charles T. Hunt

Urban Safety and Peacebuilding
New Perspectives on Sustaining Peace in the City
Edited by Achim Wennmann and Oliver Jütersonke

For more information about this series, please visit:
www.routledge.com/Studies-in-Conflict-Development-and-Peacebuilding/book-series/CONDEVPEACE

Urban Safety and Peacebuilding

New Perspectives on Sustaining Peace in the City

Edited by
Achim Wennmann and Oliver Jütersonke

Routledge
Taylor & Francis Group
LONDON AND NEW YORK

First published 2019
by Routledge
2 Park Square, Milton Park, Abingdon, Oxon OX14 4RN

and by Routledge
52 Vanderbilt Avenue, New York, NY 10017

Routledge is an imprint of the Taylor & Francis Group, an informa business

British Library Cataloguing-in-Publication Data
A catalogue record for this book is available from the British Library

Library of Congress Cataloging-in-Publication Data
Names: Wennmann, Achim, editor. | Jütersonke, Oliver, editor.
Title: Urban safety and peacebuilding : new perspectives on sustaining peace in the city / edited by Achim Wennmann and Oliver Jèutersonke.
Description: Abingdon, Oxon ; New York, NY : Routledge, 2019. | Series: Studies in conflict, development and peacebuilding | Includes bibliographical references and index.
Identifiers: LCCN 2018040900 (print) | LCCN 2018043139 (ebook) | ISBN 9781351371353 (Web PDF) | ISBN 9781351371346 (ePub) | ISBN 9781351371339 (Mobi) | ISBN 9781138554726 (hardback) | ISBN 9781315149158 (e-book)
Subjects: LCSH: Urban violence–Developing countries. | Urban warfare–Developing countries. | Public safety–Developing countries. | Peace-building–Developing countries. | Conflic management–Developing countries.
Classification: LCC HN981.V5 (ebook) | LCC HN981.V5 U73 2019 (print) | DDC 303.609173/2–dc23
LC record available at https://lccn.loc.gov/2018040900

ISBN: 978-1-138-55472-6 (hbk)
ISBN: 978-1-315-14915-8 (ebk)

Typeset in Times New Roman
by Wearset Ltd, Boldon, Tyne and Wear

MIX
Paper from responsible sources
FSC® C013056

Printed and bound in Great Britain by
TJ International Ltd, Padstow, Cornwall

Contents

Illustrations

Figures

Table

Contributors

Jonathan Luke Austin is a postdoctoral researcher at the Centre on Conflict, Development and Peacebuilding of the Graduate Institute of International and Development Studies, Geneva.

Rachel Banfield is a policy adviser in the Department of International Law and Policy at the International Committee of the Red Cross (ICRC), Geneva.

Luna Cara is an associate professor in the Faculty of Architecture and Engineering of the Universidad Internacional SEK in Quito, Ecuador.

Evaristo De Pinho Oliveira is the Head of the Water and Habitat Unit at the ICRC, Geneva.

Hugo van den Eertwegh worked with the ICRC in the field in various capacities from 2001 to 2012 before taking up work at ICRC headquarters in Geneva. He is currently Deputy Head of the Operations' Security and Crisis Management Support Unit.

Peter Homel is an adjunct professor at the Griffith Criminology Institute of Griffith University, Queensland, Australia.

Oliver Jütersonke is Head of Research at the Centre on Conflict, Development and Peacebuilding of the Graduate Institute of International and Development Studies, Geneva.

Cyntia López Rueda is an associate professor at the Faculty of Architecture and Engineering of the Universidad Internacional SEK in Quito, Ecuador.

Nicolas Masson is a human security adviser for the Federal Department of Foreign Affairs of Switzerland and also First Secretary at the Embassy of Switzerland in Lebanon.

Caroline Putman-Cramer is the former Head of Sector for the Americas and Focal Point on Youth & Urban Violence in the Central Tracing Agency and Protection Division at the ICRC, Geneva.

Tuesday Reitano is the Deputy Director of the Global Initiative against Transnational Organized Crime.

Dennis Rodgers is a research professor in the Department of Anthropology and Sociology at the Graduate Institute of International and Development Studies, Geneva, and Professor of International Development Studies at the University of Amsterdam.

Jennifer Erin Salahub managed the Safe and Inclusive Cities initiative, a global research programme jointly funded by Canada's International Development Research Centre and the United Kingdom's Department for International Development.

Mark Shaw is the Director of the Global Initiative against Transnational Organized Crime.

Hugo Slim is Head of the Policy and Humanitarian Diplomacy Division at the ICRC, Geneva.

Achim Wennmann is a senior researcher at the Centre on Conflict, Development and Peacebuilding of the Graduate Institute of International and Development Studies and Executive Coordinator of the Geneva Peacebuilding Platform.

Mayssam D. Zaaroura was a program officer working on the Safe and Inclusive Cities initiative in the Governance and Justice Program of Canada's International Development Research Centre, Ottawa. She currently works for Oxfam Canada as a women's rights knowledge specialist focused on ending violence against women and girls.

Acknowledgements

We would like to thank all the participants at the exchanges, meetings, and conferences of the Technical Working Group (TWG) on Urban Safety and Peacebuilding Practice for their reflections, inputs, and enthusiasm over the past four years. You are too numerous to thank individually, but collectively you have been a constant source of inspiration for us, and have ultimately also convinced us to pursue this edited volume.

Conceived on the margins of a major conference in Geneva in April 2013 and stimulated by the Geneva Peacebuilding Platform's annual meeting in 2013 on 'Peacebuilding in the City: Responses to Violence and Fragility in Urban Settings', the TWG built on the recognition that urban safety and peacebuilding professionals have generally worked in separate silos, even though they frequently did so in the same space: the city. After several exploratory exchanges in 2014, it became clear that there was a lot of potential to explore the confluence of urban safety and peacebuilding practice more closely. Since 2015 the TWG has brought together focal points on urban safety of cities with peacebuilding, peace mediation, and conflict prevention practitioners to help craft solutions to the rapidly increasing risk of conflict and insecurity in urban settings. The TWG had the objective of networking urban safety and peacebuilding practitioners to develop a better understanding of practice on the interface between urban safety and peacebuilding, and ultimately to contribute to developing an integrated approach to sustaining peace in the city.

The TWG evolved into a collective action initiative and counted a network of over 100 professionals by the end of 2017. It was managed as an informally convened networking initiative that mainly worked through web-based exchanges, although it held meetings in Corregidora (Querétaro, Mexico) in March 2015 and Bogotá (Colombia) in September 2015, as well as on the margins of the HABITAT III Summit in Quito (Ecuador) in October 2016. The Geneva Peacebuilding Platform also co-facilitated two conferences in collaboration with UN-Habitat and other partners on 'Reviewing the State of Safety in Cities' in June 2016 and June 2017, and incubated a discussion among Geneva-based actors about the establishment of a Geneva Cities Initiative.

The TWG would not have been possible without the constant support of Juma Assiago of UN-Habitat and the director-general of the United Nations Office at

Geneva, Michael Møller, and his colleagues Charlotte Warakaulle and Salman Bal. The conferences in 2016 and 2017 also enjoyed the support of the City of Geneva and active input from the members of a reference group representing 26 different Geneva-based institutions.

This edited volume, finally, benefited greatly from a number of anonymous reviewers chosen by the publishers, as well as from a series of experts we mobilized to offer feedback on specific contributions. We would like to thank the staff of the Graduate Institute's Centre on Conflict, Development and Peacebuilding, notably Maria-Ottavia Guicciardini and Sandra Reimann, for editorial assistance and general encouragement. Our gratitude also goes to Andrew Humphrys and his team at Routledge for their patience and support for this volume, with special thanks to Cherry Ekins for her meticulous copyediting.

The perspectives and conclusions presented in this book are those of the respective chapter authors and do not necessarily reflect the views of UN-Habitat, the United Nations Office at Geneva, or the Geneva Peacebuilding Platform. Needless to say, we bear the burden for all errors and omissions of fact or judgement.

Achim Wennmann and Oliver Jütersonke

Abbreviations and acronyms

AIC	Australian Institute of Criminology
CAG	criminal armed group
CPLC	Citizen–Police Liaison Committee
CPTED	crime prevention through environmental design
DCAF	Geneva Centre for the Democratic Control of Armed Forces
ETA	Euskadi Ta Akatasuna (Spain)
EU	European Union
FARC–EP	Fuerzas Armadas Revolucionarias de Colombia – Ejercito del Pueblo (Colombia)
FNA	National Anti-Extortion Force (Honduras)
GDP	gross domestic product
ICRC	International Committee of the Red Cross
IDF	Israel Defence Force
IDPs	internally displaced people
IHL	international humanitarian law
IHRL	international human rights law
IRA	Irish Republican Army (Ireland)
LAPOP	Proyecto de Opinion Publica de America Latina
MHS	mutually hurting stalemate
MQM	Muttahadi Qaumi Movement (Pakistan)
NGO	non-governmental organization
NSAG	non-state armed group
NTC	National Transitional Council (Libya)
NUA	New Urban Agenda
OECD	Organisation for Economic Co-operation and Development
OPT	Occupied Palestinian Territories
PAC	People's Aman Committee (Pakistan)
SDG	Sustainable Development Goal
SSR	security sector reform
TRC	Truth and Reconciliation Commission (South Africa)
TWG	Technical Working Group
UK	United Kingdom
UN	United Nations
US	United States

Introduction

New perspectives on urban safety and peacebuilding

Achim Wennmann and Oliver Jütersonke

With more than 50 per cent of the world's population now living in urban areas, and amid projections that this figure will rise to 66 per cent by 2050, cities are increasingly becoming the focal points for peacebuilding and statebuilding activities, and the primary site for the promotion of sustainable development. Demographic pressures and urbanization trends suggest an additional 2.5 billion people will reside in cities by 2050, with 90 per cent of this increase concentrated in Asia and Africa. Moreover, the most rapidly urbanizing spaces will not be megacities like Rio de Janeiro, Lagos, or Beijing, which tend to enjoy most of the attention, but medium-sized cities with less than one million inhabitants (UNDESA 2014: 1). For Africa alone, the urban population will increase by a staggering half a billion people by 2035 – the combined populations of Nigeria, Ethiopia, Egypt, and Tanzania (Bello-Schünemann *et al.* 2017: 26).

It does not take a wizard to project that these demographic trends will have profound implications on political, economic, and social orders for generations to come. On the one hand urban areas are socioeconomic magnets, sites of innovation and propinquity, and a source of perceived opportunity for many. On the other they are spaces marked by poverty, inequality, and marginalization – functionally understood in terms of a lack of access to services, including basics such as housing, education, healthcare, food, transport, security, and justice. What is more, as urbanization runs its course, many vulnerable groups may find themselves 'squeezed between declining villages and unwelcoming urban centres' (SDC and IDS 2017: 1).

The merits of close proximity and collective action are easy to overlook in such urban areas, and this volume aims to reflect on the conceptual and practical advances in the areas of urban safety and peacebuilding to help address what is viewed as being the rapidly increasing risk of conflict and insecurity in cities – with the caveat that empirically observed levels of violence and individual (and group) perceptions of insecurity often do not go hand in hand. But high rates of chronic violence are, beyond doubt, a stark reality in many cities globally. What is more, the factors driving violence and insecurity in urban areas, and insights into what can be done to reduce and mitigate such tendencies, are tackled by a plethora of academic fields and communities of practices that, arguably, do not engage with one another in a systemic manner. This, at least, is the starting claim for this volume; inviting the reader to engage in a conversation on *how* to sustain peace in the city and *who* should be part of that conversation is the aim of the nine chapters that follow.

Origins of the technical working group

With cities seen as future flashpoints, finding better responses to a changing landscape of conflict and violence is becoming an increasing priority in international security, development, and peacebuilding circles. Cockayne *et al.* (2017: 2), for instance, underline that 'the future of violent conflict is urban – because the future of humanity is urban. If we want to prevent future violent conflict, we must prevent violent urban conflict.' Yet peacebuilders, whose work tends to be associated with larger-scale armed conflicts that entail external 'intervention' of one sort or another, have only recently begun to reflect on how changing global dynamics of violent conflict also require rethinking one's conceptual categories (Krause 2014; WHO 2014; OAS 2015). While wars in Syria, Iraq, or Yemen make the headlines, the great majority of the world's violent deaths occur in 'non-conflict' settings, with Latin America, the Caribbean, and parts of sub-Saharan Africa being among the worst-affected regions (Geneva Declaration Secretariat 2015: 52). Much of this violence occurs in spatially confined areas of cities, and many urban security policies have emphasized repressive approaches where state authorities seek to crush crime through 'law-and-order' initiatives or under the banner of a 'war on drugs'. These approaches, however, have largely proven to be ineffective in terms of violence and crime reduction. In fact, heavy-handed policies and securitized responses to crime and violence have tremendous human cost and tend to exacerbate levels of violence, rather than alleviate them. While securitized approaches remain popular among many politicians, other *local* peacebuilding approaches that seek to work with affected populations in specific communities to achieve lasting solutions are now beginning to find increasing traction (Jütersonke *et al.* 2009; Eavis 2011).

At the same, the reduction and prevention of (criminal) violence have also been on the agenda of urban planners and urban designers – and perhaps to a lesser extent that of architects. But in many cities affected by high rates of violence, security policies are rarely aligned with innovative solutions to reshape public and semi-public spaces in a way that complements violence reduction strategies. On the contrary, in cities characterized by an increasing disconnect between rich and poor, infrastructure and housing projects tend to *fragment* public space further rather than reinforce it. The resulting 'architecture of fear' (Agbola 1997), driven by property developers and egged on by private security providers, is in stark contrast to the vision of 'safe and inclusive' cities propagated by UN-Habitat's Safer Cities Programme and other like-minded international agencies. These stakeholders have sought to counter *mano dura* policies by developing specific 'people first' tools and strategies to address urban vulnerabilities, crime, delinquency, and antisocial behaviours (UN-Habitat and GPP

2014) in efforts to pursue multidimensional and cross-sectoral approaches to citizen security and social integration (e.g. UNDP 2013; UNODC 2016).

Against this backdrop, the Geneva Peacebuilding Platform, in collaboration with UN-Habitat and with the support of the director-general's office of the United Nations Office at Geneva, established the Technical Working Group (TWG) on the Confluence of Urban Safety and Peacebuilding Practice in January 2015. The TWG brought together over 100 urban safety focal points and practitioners from peacebuilding, peace mediation, and conflict prevention circles in an attempt to kick-start a conversation across the perceived disconnect between localized peacebuilding efforts on the one hand and urban planning and design strategies on the other. Through a series of regional meetings and an online resource depository, the TWG fed into the lead-up discussions for HABITAT III and the New Urban Agenda (UN-Habitat 2016). That agenda explicitly sought to 'give the city back to its residents' – and thus to work against the increasing marginalization and exclusion of large parts of the urban population that is propagated by spatial strategies catering to the well-being of the privileged few. Meaningful public space, in turn, erodes further as 'containment' tactics to combat the perceived sources of 'insecurity' widen the gap – both spatial and social – between the upper echelons of society and ordinary citizens.

Sustaining peace in the city

This volume features a number of academic and practitioner contributions that have fed into the TWG process over the past three years. The book is not a summary of the TWG, nor is it the final word on the subject; rather, the contributions have been chosen for their diversity across urban safety and peacebuilding perspectives. It aims to expand this conceptual and practical interface under the banner of *sustaining peace in the city*. The notion of 'sustaining peace' stems from the recent review of the United Nations (UN) peacebuilding architecture – as documented in UN Security Council Resolution 2282 (UNSC 2016) – and repositions peacebuilding as an activity across all stages of conflict, as a priority for the entire UN system, and as a responsibility for national governments and other national and substate stakeholders. Echoing the vocabulary of the 2030 Agenda for Sustainable Development (UNGA 2015), the call to 'sustain peace' aims to support the UN's reform process and drive activities towards peaceful and inclusive societies as an explicit policy objective for all UN member states – irrespective of whether or not they are themselves confronting violent conflict. Sustaining peace involves building an infrastructure of peace-enhancing institutions, norms, attitudes, and capacities spanning different sectors and levels of social organization. It also requires an attitude of constant innovation, and is advanced through national ownership and leadership (Mahmoud and Makoond 2017; Tschirgi and de Coning 2018).

This repositioning of peacebuilding practice is an opportunity to broaden the optic of peacebuilding beyond interstate and intrastate armed conflicts – and especially their aftermath – and reconnect to the community-level origins of building peace. A nod to *sustaining peace in the city*, therefore, serves as a guiding frame within which to situate the contributions to this volume as entry points for new policy responses against conflict, violence, and exclusion in urban areas, and to promote a conversation across disciplinary and specialist silos – including architecture, criminology, social anthropology, political science, and urban studies. By exploring cross-cutting issues and a variety of urban contexts, the volume seeks to stimulate reflection about possible responses to conflict and insecurity for the coming urban century.

Jonathan Luke Austin and Oliver Jütersonke begin by clarifying the commonalities, differences, and intersections between urban safety and peacebuilding practice. They underline that urban safety and peacebuilding approaches have the potential to be mutually complementary in the fight against violence in the city, and explore how the symbiotic relationship between the public sphere of urban government (*urban safety*) and the generally more informal processes of conflict prevention, mitigation, and transformation (*peacebuilding*) can be harnessed in conjunction, rather than in contradistinction to each other. Building on insights from semiotics, the chapter seeks to offer minimalist ideal-type definitions of urban safety and peacebuilding – conceptualizations that also benefit from the medical differentiation between curative and palliative care. The chapter then constructs a heuristic typology of urban safety and peacebuilding initiatives vis-à-vis their levels of analysis, models of intervention, institutional foci, and the problems they address, before ending on a discussion of the level of *aggregation* (or granularity) of urban interventions and the level of *politicization* to which such activities are subject.

In the next chapter, Cyntia López Rueda and Luna Cara provide architectural perspectives on urban safety and crime prevention based on two case studies from the cities of Quito and Guayaquil in Ecuador. They demonstrate how the complexity of the city is characterized by the continuous change and transformation of a set of spatial configurations that determine different models and identities of urban life. These models regulate the recognition of spatial hierarchies and how they affect transitions between the public and private spheres. By studying Quito's 24 de Mayo Boulevard and Guayaquil's Simón Bolívar Waterfront, the chapter develops a conceptual framework that differentiates between *defence for safety* and *psychological safety* in terms of urban design strategies, and compares and contrasts *encounter* and *enclosure* models to offer a link between architectural typologies and urban safety policies.

Staying in the Latin American context, anthropologist Dennis Rodgers offers an alternative conception of 'gangland' that highlights its variegated nature as well as that of gang authority more generally – and in doing so, identifies potential peacebuilding avenues within what at first glance would seem to be a space whose dynamics are destructively (rather than constitutively) violent. Drawing on ethnographic research carried out in *barrio* Luis Fanor Hernández, a poor neighbourhood in Managua, Nicaragua, Rodgers describes how gangs promote a specific form of local socio-spatial ordering, one that involves places (in this case the thriving Huembes market) frequented by members of rival gangs in what the author argues is a form of *terra nullius*: a space of exception within the broader gangland context where different violent actors can coexist and engage with each other in ways that do not have to entail antagonism. Such spaces are important, because they open possibilities for concrete trust-building initiatives within gangland areas.

Taking a broader view on what he called criminal armed groups (CAGs), Hugo van den Eertwegh develops an argument about integrating negotiations with such actors into the violence reduction toolbox. Analysing *when* and *how* to negotiate with such groups, the chapter builds on the recognition that armed violence is a major threat to many different societies and finding solutions for it requires recourse to the full range of policy options, including negotiation strategies with CAGs and building the technical skills to conduct such negotiations. Van den Eertwegh argues for innovative and inclusive forms of citizen participation to strengthen the democratic foundations of negotiations with CAGs and thus create sustainable negotiation outcomes in – and with – empowered communities.

Tuesday Reitano and Mark Shaw explore the growth and operation of extortion economies in three urban contexts – Central America, Karachi, and Cape Town – to show how they are critical in shaping and sustaining criminal forms of governance in cities, and in particular in post-conflict and high-violence localities. Given that available policy options to counter extortion-driven forms of governance have not been discussed in detail outside a few contexts (notably Italy), the chapter discusses whether there are lessons to be learned from these approaches for peacebuilding and the establishment of legitimate forms of urban security management.

Returning specifically to the conceptual and practical interface between urban safety and peacebuilding, Jennifer Salahub and Mayssam Zaaroura look at the opportunities for gender analyses to strengthen the nexus, asking specifically how one might go about integrating a gender lens into practical work to help improve safety for both women and men in urban centres. They argue that much of the work on urban safety and peacebuilding remains gender blind, outlining the missed opportunities, decreased operational ineffectiveness, and existing exclusive and exclusionary power structures that are reinforced as a result. The chapter builds on a review of emerging good practice and recommendations for practical changes, and draws on research from Indonesia, Egypt, and Libya.

Peter Homel and Nicolas Masson reflect on the intersection between community crime prevention and security sector reform (SSR). Based on their experience from SSR initiatives undertaken in the Jenin Governorate in the Occupied Palestinian Territories, the chapter highlights a range of shortcomings found in traditional top-down SSR approaches and contrasts them with an alternative, bottom-up, community-based approach that uses outcome-based local crime prevention planning processes to build effective 'partnerships' for crime prevention with both formal (e.g. security forces, executive authorities, parliamentarians, governors' offices) and informal (civil society, media, tribal and business leaders) security providers. The chapter highlights how such processes constituted a viable mechanism by which to create a safer community with stronger local leadership.

Hugo Slim, Caroline Putman-Cramer, Evaristo De Pinho Oliveira, and Rachel Banfield look at how the International Committee of the Red Cross (ICRC) supports authorities in cities affected by war and violence to maintain access to essential services such as safe water, safe health facilities, and safe schools. They draw on the ICRC's operational experience in armed conflict in the Middle East, Africa, and Asia, as well as the organization's work in settings of urban violence in Latin America. The chapter examines the dynamics of danger and risk for vulnerable populations in different urban areas, and illustrates how the ICRC and its partners in local authorities and national Red Cross and Red Crescent societies work to limit these risks and maintain essential services.

In the concluding chapter, Achim Wennmann draws together the main themes of the volume and reflects on the *how* in sustaining peace. By emphasizing that the confluence of urban safety and peacebuilding practice can result from the practical need to solve problems in specific places, the chapter explores the options for achieving safer and more inclusive cities. It shows how both urban safety and peacebuilding professions can look back at a long record of driving participatory processes and working within contentious political economies. The degree of local agency for change points to the existence of spaces that frame and provide oxygen to transformative processes aiming to reduce violence and build peace in cities. These spaces exist in different shapes and forms, but there is much expertise on how to create and maintain them as a means to drive negotiated change. This operational focus brings into the fold a series of unconventional actors, processes, and ways of working that evolve from the alignment of partisan interests on the pathways for peace. By garnering know-how about working politically, urban safety and peacebuilding professionals have much to contribute to the co-production of safer and inclusive cities.

Given the current state of safety in cities and the future risks of increased turbulence, there is much work to do and there are many places to start. The analysis and practice presented in this volume should remind those willing to work towards peaceful and inclusive cities that there are tried and tested approaches available, and a host of experts and practitioners ready to accompany those prepared to lead in their respective contexts.

References

Agbola, T. 1997. *Architecture of Fear: Urban Design and Construction Response to Urban Violence in Lagos, Nigeria*. Ibadan: African Book Publishers.

Bello-Schünemann, J., J. Cilier, Z. Donnenfeld, and C. Aucoin. 2017. *African Futures: Key Trends*. Pretoria: Institute for Security Studies.

Cockayne, J., L. Bosetti, and N. Hussain. 2017. 'Preventing Urban Conflict', briefing paper. Tokyo: United Nations University Centre for Policy Research.

Eavis, P. 2011. *Working against Violence: Promising Practices in Armed Violence Reduction and Prevention*. Geneva: Geneva Declaration Secretariat.

Geneva Declaration Secretariat. 2015. *Global Burden of Armed Violence: Every Body Counts*. Geneva: Geneva Declaration Secretariat.

Jütersonke, O, R. Muggah, and D. Rodgers. 2009. 'Gangs, Urban Violence, and Security Interventions in Central America', *Security Dialogue* 40(4–5), 373–397.

Krause, K. 2014. 'Critical Perspectives on Human Security', in M. Martin and T. Owen (eds) *Routledge Handbook of Human Security*. London: Routledge, 76–93.

Mahmoud, Y. and A. Makoond. 2017. *Sustaining Peace: What Does It Mean in Practice?* New York: International Peace Institute.

OAS (Organization of American States). 2015. *Searching for Common Approaches to Deal with Unconventional Conflicts and Violence in the Americas*. Washington, DC: OAS.

SDC and IDS (Swiss Agency for Development and Cooperation and Institute for Development Studies). 2017. 'Cities, Urbanization and Poverty Reduction', briefing paper, available at www.shareweb.ch/site/Poverty-Wellbeing/currentpovertyissues/Documents/IDS%20SDC%20Urban%20Poverty%20Brief%20final.pdf (accessed 5 July 2018).

Tschirgi, N. and C. de Coning. 2018. 'The Challenge of Sustaining Peace: Enhancing and Moving Beyond the United Nations' Peacebuilding Architecture', in W. Durch, J. Larik, and R. Ponzio (eds) *Just Security in an Undergoverned World*. Oxford: Oxford University Press, available at www.oxfordscholarship.com/view/10.1093/oso/9780198805373.001.0001/oso-9780198805373-chapter-17?print=pdf (accessed 27 June 2018).

UNDESA (UN Department for Economic and Social Affairs). 2014. *World Urban Prospects*. New York: United Nations.

UNDP (UN Development Programme). 2013. *Citizen Security with a Human Face: Evidence and Proposals for Latin America*. New York: UNDP.

UNGA (UN General Assembly). 2015. 'Transforming Our World: The 2030 Agenda for Sustainable Development', UN Doc. A/RES/70/1, 21 October. New York: United Nations.

UN-Habitat (UN Human Settlements Programme). 2016. *The New Urban Agenda*. Nairobi: UN-Habitat.

UN-Habitat (UN Human Settlements Programme) and Geneva Peacebuilding Platform. 2014. 'City Labs for Safer Cities: Towards an Integrated Approach to Urban Safety and Peacebuilding', consultation draft, 24 March. Nairobi and Geneva: UN-Habitat and Geneva Peacebuilding Platform.

UNODC (UN Office on Drugs and Crime). 2016. *Governing Safer Cities: Strategies for a Globalized World*. Vienna: UNODC.

UNSC (UN Security Council). 2016. 'Resolution 2282 on Post-Conflict Peacebuilding', UN Doc. S/RES/2282, 27 April. New York: United Nations.

WHO (World Health Organization). 2014. *Global Status Report on Violence Prevention*. Geneva: WHO.

1 Understanding the grammar of the city

Urban safety and peacebuilding practice through a semiotic lens

Jonathan Luke Austin and Oliver Jütersonke

Introduction

The relationship between urban safety, peacebuilding practice, and efforts to sustain urban peace is both obvious and obscure. It is obvious in the mutual focus of practitioners working within these fields on the problems of crime, violence, and societal conflict in urban centres. It is obscure when it comes to disentangling what is actually distinct about the practical or analytical content of the programming of each approach in dealing with these issues – not least because their similarities often result in a great deal of overlap when it comes to practical activities. The *difference* between the two approaches, not the overlap, is what needs clarification today. In this chapter we attempt to provide this clarification by exploring the commonalities, contrasts, and intersections between urban safety and peacebuilding practice.

Providing greater clarity over the distinction between the two approaches is important because, in the absence of conceptual precision, resources risk being wasted on redundant programming instead of fostering a constructive complementarity harnessing the expertise of each approach. The politically charged environment in which this debate is carried out – and to which this entire volume attests – only intensifies the risk. Because each approach involves different communities of practice (informed by different academic and expert circles), as well as potentially competing donor and state interests, debates over the nature of the relationship between urban safety and peacebuilding are not just technical but also about political interests and priorities. More than this, developing an integrated approach that leverages the expertise of both is critical because today's urban centres, according to Chinese scholars Kaizhen Cai and Jinguo Wang (2009: 219), face a 'complex and urgent situation' that goes beyond 'traditional safety problems' like traffic accidents, natural disasters, and crime incidents to involve 'nontraditional security problems' such as terrorism, pandemics, and eco-environmental safety.

Indeed, the contemporary challenges faced by urban safety and peacebuilding practitioners relate to a complex combination of different types of harms occurring in cities, which are settings whose conflictual dynamics are still relatively poorly understood (see Jütersonke and Krause 2013; Jütersonke *et al.* 2007). As we explore in this chapter, the texts in this volume all broadly take the view that urban safety and peacebuilding have the potential to be mutually complementary in the fight against this complex set of urban challenges. To achieve this, however, we need to clarify their distinct areas of expertise by asking how the symbiotic relationship between the public sphere of urban government (urban safety) and the generally more informal processes of conflict prevention, mitigation, and transformation (peacebuilding) might be harnessed. To answer this question, we need to understand the particular expertise of the urban safety approach and how it contrasts with and complements that of peacebuilding. In what follows we seek to pursue that task, proceeding in five main stages.

First, we define and clarify the concepts of urban safety and peacebuilding in *ideal-type* and *minimalist* terms. We parse back to the basic roots of the concepts to see how their genealogy (i.e. the evolution of their meanings over time) may affect the multiplicity of definitions and understandings that can be gleaned from contemporary practice. Second, with minimalist definitions in hand, we dissect the productivity of the interface between urban safety and peacebuilding by introducing a basic model of their approaches inspired by *semiotics*: a field of study looking at the way in which 'signs' shape the way we make sense of everyday interactions by exploring how our use of language and interaction with material objects (e.g. traffic signage) construct human society. This model clarifies how each approach *in its minimalist ideal type* takes a different route towards 'meaning-making' in urban settings so as to reduce the possibility of crime, violence, and societal conflict. We then construct a heuristic typology of the foci of urban safety and peacebuilding initiatives vis-à-vis their levels of analysis, models of intervention, institutional foci, and the problems with which they deal.

Third, from this conceptual base the chapter draws a more practical analogy between the two modes of meaning-making outlined by examining two forms of care that constitute the basis of medical interventions: *curative* care and *palliative* care. We argue that peacebuilding, conceived in its minimalist ideal type, seeks to 'cure' violence; by contrast, urban safety has usually been concerned with mitigating and reducing harm. And while these divisions have blurred over time as the two approaches have become intertwined, they remain driving forces that tend to be hidden in the shadows of ongoing programming discussions. Fourth, we continue by dealing with the key problem faced by the minimalist and idea-type description of the contrasts between the two approaches: namely that they are frequently employed together and often 'drift' towards one another. We clarify that both urban safety and peacebuilding initiatives can be analysed in terms of the level of *aggregation* (or granularity) of their interventions, as well as the level of *politicization* to which their activities are subject. We discuss several pressures that have forced both approaches to deviate from their ideal-type form, beyond simple programmatic necessity, and chart their resulting confluence schematically.

Fifth and finally, we conclude with a discussion of how our semiotic clarification of the relationship between urban safety and peacebuilding practice can assist us in moving towards a truly integrated approach. This entails highlighting dangers in their confluence as well as opportunities that will enable us to leverage their respective strengths. The conclusion therefore stresses how it may be useful to parse back to the minimalist and ideal-type forms of urban safety and peacebuilding initiatives to see how this genealogy of their origins continues to impact how they are being implemented today, even as the scope of their initiatives has broadened and blurred considerably. Doing so, we suggest, will allow a better division of labour between the two approaches moving forward.

Minimalist conceptualizations of urban safety and peacebuilding

Twenty years after Boutros Boutros-Ghali's now famous Agenda for Peace (UNSG 1992) laid out a basic framework on which peacebuilding practice has to proceed, we must begin by noting that:

> ... *the post-conflict framing of peacebuilding has been overtaken by the evolution of violent conflict and the ensuing adaptation of the response.* Peacebuilding practice now occurs in a wider variety of contexts ranging from fluid political transitions to regions under increased stress due to climate change, rapid urbanization, or contentious large-scale investments.
> (Geneva Peacebuilding Platform 2015a: 3, emphasis added)

Against the backdrop of such a fluid and complex programming context, it has become clear that generating a one-size-fits-all conceptualization of peacebuilding has become increasingly futile and even counterproductive (Geneva Peacebuilding Platform 2015b). Instead, practitioners and scholars working within the domain of peacebuilding have come to focus on what are thought to be some of the essential characteristics of efforts to build peace, including 'the use of dialogue, trust-building, and consensus-seeking processes to resolve or manage conflict through non-violent means' (ibid.: 6). Such an approach is taken in acknowledgement of the fact that 'at the field level, many people simply get on with doing what is needed to build peace and do not worry about definitions' and, in this vein, it seems that analytical accounts of peacebuilding might also work without 'preconceived analytical categories (e.g. a particular definition of peacebuilding)', instead allowing 'patterns of consent and disagreement to emerge from the data itself' (ibid.: 17).

Making sense of peacebuilding in this adaptive manner has proved crucial to keeping up with a rapidly changing world. However, it is also one of the main difficulties obscuring the nature of the relationship between urban safety and peacebuilding practice. We believe that clarifying this relationship requires an *ideal-type return to minimalist definitions of peacebuilding and urban safety*. We do not mean to suggest these definitions are in any way preferable to current understandings, or in any way reflect the reality of practice today. Quite the contrary: broader contemporary definitions are the basis of countless productive programming initiatives globally, and are continuously debated and redrawn in both multilateral forums such as the United Nations – reflected in the recent turn to the vocabulary of 'sustaining peace' in United Nations (UN) Security Council Resolution 2282 (UNSC 2016) – and community-driven peacebuilding work at the grassroots level. We are seeking to return to minimalist definitions only in order to appreciate the conceptual history of these broader understandings of urban safety and peacebuilding. We argue that the origin of current approaches in these minimalist definitions continues to affect many of their activities, and appreciating this is thus important to clarify the relationship between urban safety and peacebuilding practice today.

We begin, then, by outlining our ideal-type minimalist understandings of urban safety and peacebuilding. The term 'urban safety' is relatively specialized in comparison to 'peacebuilding'. Indeed, charting the use of these terms between 1970 and 2005 based on Google Ngram data from collated printed material reveals the vast rise in the use of peacebuilding terminologies over the 1990s and – by contrast – the almost imperceptibly low use of urban safety as a term of reference across the same period. Reviewing the predominance of peacebuilding in this way is misleading, however. Urban safety programmes are most frequently subsumed within urban *planning* schemes, and UN-Habitat's Safer Cities Programme itself discusses 'enhancing safety through *planning*, management and governance' (UN-Habitat 2012; emphasis added). Moreover, safety and security are often conflated in these debates – sometimes, indeed, because of linguistic particularities (e.g. safety and security are both termed *seguridad* in Spanish, as discussed by López Rueda and Cara in this volume) that tend to occlude the distinction. What is clear, however, is that urban safety is closely related to the field of urban planning (and other subfields therein, including urban ecology, urban security, urban design, etc.). Indeed, urban planning is a practical and disciplinary perspective whose work long *precedes* the rise of peacebuilding (for reviews see Levy 2016 and Kostof 1991), despite the two terms being employed with almost the same degree of frequency today.

The very development of the first cities and their planning was, indeed, about the safety they afforded: 'the city was a more defensible place than an isolated settlement in the countryside' (Levy 2016: 62). However, as John M. Levy explains, this early virtue of the city gradually decreased with technological innovations like cannons that could breach city walls, nuclear weapons that could destroy whole cities, and terrorists drawing on everyday objects to penetrate and cause harm inside the city itself (ibid.: 62–63). In combination with the more mundane problems of crime and societal conflict, urban planning became increasingly concerned with the security or safety of the city more generally. A comprehensive urban planning approach is thus, Levy suggests, one that encompasses *health, safety*, and *public welfare*. The specifically urban *safety* component of these three elements, he continues, can 'manifest itself in numerous' different ways:

> It might mean requiring sufficient road width in new subdivisions to ensure that ambulances and fire equipment have adequate access in emergencies.

> Many communities have flood plain zoning to keep people from building in flood-prone areas. At the neighbourhood level it might mean planning for a street geometry that permits children to walk from home to school without crossing a major thoroughfare. In a high-crime area it might mean laying out patterns of buildings and spaces that provide fewer sites where muggings and robberies can be committed unobserved.
>
> (Ibid.: 122)

Whereas other aspects of the comprehensive urban planning approach (those relating to health and public welfare) focus more directly on intervening at a human level in the activities of urban dwellers, what is most striking about the urban safety aspect is its focus on *material* structures. It is by altering the material environment that urban safety hopes to reduce the possible hazards of the city, and it is important to keep this ideal-type definition of urban safety in mind as our discussion proceeds.

The definition of peacebuilding, for its part, has varied over time as it was regularly revised and *expanded* in scope. Numerous definitions and conceptualizations of peacebuilding now abound (see Barnett *et al.* 2007; also Chetail and Jütersonke 2014 for an overview of the literature). Building on the work of Tadashi Iwami (2016), variations in our understanding of peacebuilding can usefully be categorized with respect to four substantive elements: the primacy of the military; the importance of order and stability; engagement with political and democratic processes; and state-building agendas and infrastructure development. According to Iwami, the military and related 'stabilization' ambitions play a particularly important role in the American conceptualization of peacebuilding as a tool fostering democratic transformation (e.g. post-war Afghanistan/Iraq), while the European approach to peacebuilding places a premium on good governance and political reform agendas that do not necessarily depend on (military) security concerns and are focused more on interventions on the community level. For the sake of comparison, Russian and Chinese understandings of peacebuilding appear to be quite similar to the American variant but are distinctive in the sense that they downplay the importance of domestic transformation: the *mission civilisatrice* (Paris 2002) of so-called 'liberal peace' agendas on democratization, good governance, and human rights promotion (see e.g. Richmond 2006; Heathershaw 2008; Debiel *et al.* 2016) that are seen as interference in domestic affairs by some (Stepanova 2004; Lei 2011) and championed by others.

In this chapter, however, we deliberately try to steer clear of this ideologically and bureaucratically motivated terminological minefield; instead, we propose to return to the more minimal intellectual genesis of the field. Arguably, that genesis lies in the work of Norwegian sociologist Johan Galtung (1976), who programmatically outlined three approaches to peace: the dissociative approach (in which interventions aim to avoid being directly involved in conflicts, as is most prominent in peacekeeping missions), the conflict resolution approach (in which 'root causes' of conflicts are tackled), and the associative approach (where interventions are embedded and actively involved in conflict dynamics). It was perhaps a combination of Galtung's second and third approaches that first came to dominate contemporary peacebuilding practice, and was popularized by Boutros Boutros-Ghali's 1992 Agenda for Peace as 'action to identify and support structures, which will tend to strengthen and solidify peace in order to avoid a relapse into conflict' (UNSG 1992). Of course, debate has moved forward considerably in the intervening years, and this chapter does not propose to legitimate what some might critically label a 'standard UN' definition of peacebuilding. What we are interested in is drawing out some of the essential elements of a *minimalist* understanding of peacebuilding derived by returning to the early agendas and debates of the 1990s, and using this hereafter in contrast with urban safety approaches so as to make the nature of the relationship and (the possibility of) the confluence of the two less obscure.

Before discussing the relationship between urban safety and peacebuilding, we must note one central point: there is, in fact, no *necessary* relationship between the two approaches. That is, urban safety programmes do and should exist in the absence of peacebuilding schemes in many states that do *not* suffer from high levels of crime, violence, and societal conflict. Urban safety schemes exist in Switzerland, but peacebuilding schemes do not. Indeed, comprehensive urban safety measures will be necessary in any city and at any stage of a country's urbanization. Of course, the particular interest of this volume lies is in their interconnection when both are perceived to be necessary (or at least complementary) for a variety of reasons. But to appreciate the potential merits of thinking about the interface between urban safety and peacebuilding, it is important to keep in mind that urban safety schemes were designed for and are still most commonly used in cities whose contexts are quite distinct from those more usually focused on in peacebuilding practice.

Meaning-making in the city: a semiotic clarification

We now schematize an *ideal-type* relationship between the *minimalist* understandings of urban safety and peacebuilding sketched above. We do so by drawing on the tools of semiotics. Semiotics is an academic approach to studying everyday 'meaning-making' processes (for a basic introduction see Sebok 2001). Put most simply, semiotics demonstrates how the meaning of one 'sign' (say, the word *safety*) is only meaningful in relation to other words (say, the word *violence*). It therefore studies how words or other signs are ordered in sequence with each other at different levels. As a field of research, semiotics has usually been preoccupied with studying the use of language by human beings, but also extends to a broader understanding of signs as involving both written or spoken language, visual languages, and material objects that communicate meaning. It is thus possible to draw a distinction between classic semiotics and a more contemporary 'material semiotics' that integrates the material side of the world into systems of meaning-making, as opposed to seeing the world as contained solely within the minds of human beings (on material semiotics see, particularly, Austin 2016).

Take the example of a stop sign on a road. It has a textual element – the written word STOP, usually in block capitals – but also visual elements, including the block capitals themselves, the sign frequently being painted red, and its other qualities as a material object (its particular shape, height, visibility, etc.). All these aspects affect how drivers perceive the sign (how easily they see it, notice it, pay attention to it) and so work to 'make meaning' in the interaction between the driver and the road. Indeed, just such meaning-making activities are central to a very basic type of urban safety programme involving these same road signs. UN-Habitat, for example, notes how one effect of urbanization is an increase in road traffic accidents, which:

> ... result from a combination of structural, physical and behavioural factors. While the exposure of road users to traffic accidents is shaped by physical aspects of the road environment, individual behaviour, awareness of safety regulations and travel habits also determine vulnerability to traffic accident risks. In addition, the safety and design features of vehicles shape the likelihood of being involved in a traffic accident, as well as the severity of the impact.
>
> (UN-Habitat 2007: 221)

Consider, moreover, this full example of a:

> 4 A.M. [car] crash that was classified by police as caused by a drowsy driver. Yes, if the driver in question did not drive past his or her 'bedtime' (driver factor) the crash would not have occurred. However, the crash could have also been prevented by a drowsy-driver detection system (a vehicular factor), a road-departure warning system (a vehicular factor), or an effective rumble strip that alerts the driver if leaving the lane (environmental factor).
>
> (Sivak and Tsimhoni 2008: 456)

In these examples we find at least *two* broad ways in which to prevent people from coming to harm; two ways to make them 'safe'. The first involves focusing on the behaviour of human beings themselves, who may cause a situation to become unsafe because of their actions – when driving, for instance, they may be fatigued, intoxicated, or otherwise incapacitated. To stop that from happening it is necessary to alter the motivations or consciousness of people. This can be achieved through various means, but most usually it involves a combination of educating people about safe behaviours such that these safe practices become habitual, *and* enforcing these acceptable behaviours through legal sanctions and punishments that deter deviation from them (e.g. making it illegal to drive when intoxicated). In doing so, it is hoped that the 'root causes' of the harmful action can be minimized.

The second way to increase safety involves dealing with the *urban fabric* of a setting, which can be defined as 'the social and physical differentiation of urban space and its related social processes', or, as the architect Anastasia Loukaitou-Sideris expands:

> Urban fabric is composed of the material and physical aspects of the built environment in cities – the buildings, streets and alleyways, sidewalks, open spaces, and other micro-environments that represent the settings of everyday life. The *urban fabric is the container of social activity*; its layout and design can encourage or discourage certain types of behaviour.
>
> (Loukaitou-Sideris in Ceccato 2012: 6, emphasis added)

The 'vehicular' or 'environmental' factors that can prevent traffic accidents – road-departure warning systems, effective rumble strips, etc. – are part of the urban fabric that 'contains' social activity. By altering these elements of a social activity like driving, one may be able to hold the original causes of danger (fatigue, intoxication, illness, etc.) constant and yet still mitigate the potential for harm. It is thus, for example, that the terminology of urban safety often employs more neutral terms like 'hazard' as opposed to 'threat' or 'danger' (ibid.). The principle here is that the urban fabric might contain hazards that can be reordered to reduce the potential for harm, irrespective of *deliberate* human behaviour. And it is for this reason that the comprehensive urban planning approach focuses, as we discussed in the preceding section, on the *materiality* of urban safety planning in cities.

In the terms of semiotics, the two types of harm reduction mentioned here can be seen as altering either the *paradigms* or the *syntagms* of the city (Sebok 2001). A paradigm is a particular sign (word, image, or person) that can be replaced with another in a mutually exclusive (contrastive) fashion. A syntagm, by contrast, relates to the order, structure, or syntax in which these paradigms are arranged. Famously, Roland Barthes (2013) describes paradigms and syntagms in terms of clothing. Paradigms are things like jackets, shirts, or blouses that we wear one at a time; they are basic units of which different types cannot be worn simultaneously. By contrast, syntagms refer to the order or 'assembly' of clothing to alter the meaning of these basic parts. For example, a shirt which is neither tucked into the trousers nor ironed is a syntagmatic element of clothing and makes a different kind of meaning (to the wearer and the observer) to a shirt that is tucked in and well ironed.

For our purposes, the first type of harm reduction described above, which focuses on removing an original cause from a situation entirely, can be seen as *paradigmatic* in its approach. If people can be persuaded not to be violent or to commit crimes – whether that occurs through education, socioeconomic betterment, or any other means – then the whole paradigm of the city shifts. Here, meaning is made in the city by altering the very identity of the actors. People become 'better' at being 'safer'. The second type of harm mitigation is *syntagmatic*: it does not alter the original cause of harm but alters its 'syntax'. The basic elements of the situation are allowed to remain the same, with people sometimes doing things that increase the risk posed to themselves and others. What is changed, however, is the *order* with which the individual interacts with the urban fabric. The 'social container' (for their action) is modified so as to alter the meanings of being in the city in one way or another. And these reorderings are hoped to reduce the hazards of the urban fabric of the city.

For semiotics, then, paradigms and syntagms are the basic units of the organization of reality and its meaning. They refer to the concepts and *grammar* of the human world. What we want to suggest is that it might be fruitful to think similarly about different *grammars of the city* – different ways of making meaning in the city – that are more or less conducive to the emergence of crime, violence, and societal conflict. Indeed, *paradigmatic* changes to the grammar of the city are, we propose, akin to those that *minimalist peacebuilding* interventions generally seek to carry out, while *syntagmatic* changes are more aligned with what *urban safety* activities focus on. This becomes clearer if we typologically analyse the distinctive foci of the two approaches, as in Table 1.1.

Again, we stress the caveat that Table 1.1 is depicting *minimalist ideal-type* understandings of urban safety and peacebuilding. Its second and fourth columns should *not* be construed as representing the entire spectrum of peacebuilding or urban safety practices being carried out today, *nor* do they seek to describe the 'theories of change' that actual programming might follow. Generally speaking, however, ideal-type urban safety programming focuses on 'hazards' posed by the urban fabric at a very *micro* level of analysis. The focus is on the ordering of interpersonal interactions in the public sphere, between people and roads, buildings, rivers, streets, lighting systems, surveillance technologies, and so on. And there is rarely a 'top-down' means – in its ideal-type understanding – by which to achieve this. Streets are taken one at a time and are, microscopically, reordered syntagmatically. By contrast, minimalist ideal-type peacebuilding programmes are usually more *macro* in the means of their implementation. While it is more than possible to distinguish between a 'top-down' and a 'bottom-up' approach to peacebuilding (and statebuilding; see e.g. Wennmann 2010), wherein the former is more closely associated with state-led action and the latter with grassroots engagement with specific target groups, both approaches are arguably holistic in their efforts going either up or down – they seek eventually to deal with the phenomena of crime, violence, and societal conflict at a *systemic* or *structural* level.

To move away from this more technical vocabulary, a useful analogy can be drawn with types of medical care. Generally speaking, urban safety programmes do not seek to 'cure' problems of violence, conflict, or insecurity in urban settings; rather, they seek to reduce the risk posed by these problems and the possibility of their escalation. In this respect the focus of urban safety practitioners on the syntagmatic elements of the grammar of the city serves a *palliative* role. Palliative care *relieves* the symptoms of illness or, in this case, of urban crime, violence, and societal conflict. The goal of urban safety is similarly to *improve quality of life* for all those living in the city, without necessarily (but see below) working to cure the underlying problems that are causing these harms. By altering the syntagmatic ordering of the city – the situation of violence – it is hoped that the propensity for *further* or *additional* harm will reduce. In contrast, working to alter the paradigms of the grammar of the city can be seen as a form of *curative* care. Curative care seeks to tackle the underlying causes of harm, whatever they may be. This model of care aligns with the ideal-type modality of intervention that is explicitly described in minimalist peacebuilding schemes, and in particular the

Table 1.1 Ideal-type contrasts between urban safety and peacebuilding

	Minimalist understandings of urban safety	*Possible points of confluence*	*Minimalist understandings of peacebuilding*
Level of analysis	Micro *The everyday interactions of people with the urban fabric around them in the city.*	Meso *A holistic view of people's interactions with the city – materially and ideationally.*	Macro *The overall 'structure' of peoples interactions with the city, usually ideationally.*
Nature of problem	Hazards *Non-intentional consequences of human interaction with the urban fabric.*	Risk *A more-or-less intentional harm caused by the interaction of persons and the urban fabric.*	Threats *A normally intentional risk of harm being caused by persons in the city.*
Unit of institutional focus	Street *The situational contours within which human interaction occurs in public space.*	City *Representational governance and a 'right to the city' at the municipal or communal level.*	State *National government seeking to achieve sustainable peace (frequently from the local level up).*
Unit of intervention	Urban fabric *A focus on the materiality of the urban fabric of the city and its impact on persons.*	Interaction *A focus on both ideas and materiality and their dual role in a city's grammar.*	Urban dwellers *A focus on the ideas of city dwellers and their role in their interactions in the city.*
Temporal form of intervention	Constant *There is no start or stop; urban safety works constantly in the background.*	Iterative practice *Constant measures necessary to reach the goal of 'peace' in an iterative learning process.*	Linear *'Peace' can be achieved at time Y if we takes measures at time X.*
Semiotic form of intervention	^ Syntagmatic ^	^ Grey zone ^	^ Paradigmatic ^

conflict resolution and associative approaches we mentioned when defining the minimalist conceptualization of peacebuilding earlier. The aim is to address 'root causes' of violence and conflict, while embedding peacebuilding interventions within 'local' or 'indigenous' structures and capacities that are identified as crucial mechanisms for effective conflict mitigation. This, it is hoped, will lead to a *paradigmatic* shift away from crime, violence, and societal conflict in the city. The goal, indeed, is to *cure* violence (Cure Violence 2016).

Put in these terms, the distinction between urban safety and peacebuilding practice is relatively clear. More than this, it suggests the importance of clarifying the division of labour between the two approaches. Consider Johan Galtung's work *Peace by Peaceful Means*, which contains many references and comparisons to medical science, including the following:

> There is nothing wrong with schools of medicine, in the plural; *the problem is when only one school is available*. Insiders and outsiders alike are entitled to a choice. The field can benefit from diversity only as long as it is interactive in general, and dialogical in particular, with a view to mutual enrichment and in some cases to a synthesis.
>
> (Galtung 1996: 15, emphasis added)

Ideally speaking, urban safety and peacebuilding would act as different schools of medicine, palliative and curative, and mutually reinforce each other. But why, then, does general obscurity prevail?

To answer that question we need – again – to recognize the fact that most practical interventions 'in the field' of urban settings regularly draw on *both* the palliative care methods of urban safety *and* the curative methods of peacebuilding (in their ideal types) to create a hybrid form of 'peacebuilding in the city' that blurs the lines of the two approaches. Most groups carrying out activities in the field are arguably working in column three of Table 1.1 – constituting a grey zone that reflects the complexities of contemporary programming contexts. Seeking to find new intervention modalities to meet the challenges posed by this complexity, the urban safety and peacebuilding communities have moved away from the minimalist conceptualizations outlined in columns two and four of the table. 'Field realities' in cities have thus brought together both researchers and practitioners from the urban safety and peacebuilding communities by simple necessity. Nonetheless, we would argue that keeping in mind the minimalist intellectual and programmatic origins of these two approaches is important, as their basic inclinations are often still found in the broad approaches of many organizations. Appreciating these distinctions will help us foster a relationship that avoids redundant overlap and harnesses the power of the two approaches into better-integrated programming. In short, then, we suggest that minimalist peacebuilding programming deals *curatively* with the *paradigmatic* elements of the grammar of the city, while urban safety deals *palliatively* with the *syntagmatic* aspect. However, as already noted, this ideal-type distinction between the two *is not necessarily lived in practice*, as we will now discuss in more detail.

Aggregation, politicization, and urban safety and peacebuilding practice

Distinguishing between urban safety and peacebuilding measures in terms of their curative or palliative roles is important, but it remains an ideal-type distinction. In reality the lines are blurred, and this blurring of their roles is increasing. To appreciate why this is so at a conceptual level – and so judge the opportunities and challenges of this confluence – we now further distinguish between the two approaches in terms of the effects of their respective palliative or curative roles along two additional axes: *aggregation* and *politicization.*

Let us begin with *aggregation.* This refers to the degree to which social or technical practices are either analysed and intervened in on a case-by-case basis, or aggregated together. It is essentially, again, a distinction between micro and macro levels of analysis and intervention. In other words, to what degree are these approaches 'granular' or 'aggregated' in their interventions in the city? Ideally speaking, urban safety schemes intervene at a very low level of practical aggregation: roads should be studied individually – by each street corner they entail and each sidewalk that runs alongside them – to identify any hazard that might be present. By contrast, peacebuilding schemes ideally operate at quite a high level of aggregation such that they can holistically deal with 'root causes' of conflict. It does not necessarily make sense to, say, direct a peacebuilding scheme in Israel and the Occupied Palestinian Territories on a street-by-street basis. Instead, it is generally hoped that changing ideas at a broader level of societal aggregation will effect change.

Frequently, however, neither urban safety nor peacebuilding activities achieve these ideal-type positions. Urban safety schemes, for one, are often *not* particularly granular in their practices. This appears to be the case because the granularity of urban safety practice varies in relation to the material wealth of the society in which it is being enacted. Granular urban safety activities are expensive. A truly comprehensive urban safety scheme includes a focus on not only aspects of the urban fabric directly related to hazards like crime or violence but also – more broadly – issues relating to public health (e.g. building water or sewage facilities, constructing septic tanks, and separating industrial activities from residential areas), ensuring 'circulation' in the city (adequate and safe road networks, parking infrastructures, storm hazard planning and evacuation measures, etc.), the provision of public services conducive to health and safety (including parks, recreation areas, and hospitals), environmental protection (reduction of greenhouse gas emissions, building restrictions in dangerous areas, etc.), and many more (Levy 2016: 122–123). Less wealthy states, particularly those situated in the global South, are unlikely to possess the resources to be able to implement all these measures at an adequate level of granularity, and this is especially true in conflict-affected societies.

By contrast, in protracted crises peacebuilding schemes often become stuck in their capacity to act at a high level of aggregation that might shift the semiotic *paradigm* of a conflict. Thus many peacebuilding schemes descend in their level of aggregation – by necessity – to dealing less with 'root causes' and more and more with the case-by-case activities of urban safety practitioners. In other words, they shift from curative to palliative care giving. Take again the example of the Israel–Palestine conflict and the UN Relief and Works Agency and other UN agencies that are nominally working towards 'peace'. Increasingly, these agencies operate as a 'state within a state' that provides basic social services, including urban safety measures, to alleviate poverty, crime, and suffering in the increasingly urbanized West Bank and Gaza Strip (see Bocco and Takkenberg 2009). A clear shift, then, from curative to palliative medicine.

Civil society organizations in the West Bank have been similarly afflicted. Tariq Dana (2013) argues that Palestinian societal groups have become increasingly depoliticized as hope of negotiating a peace agreement has floundered and external pressures have forced their hands. The result has been a focus on increasingly granular 'technical' solutions that alleviate immediate suffering. As opposed to being focused on altering the paradigm of the conflict at quite a high level of aggregation (e.g. changing broader international ideas about the conflict and/or societal norms that underlie it on both sides), the groups have shifted to the everyday work of dealing with the urban fabric of the city and reducing harm, rather than building peace *per se*. Again, we can see in this disaggregation of the approach a clear shift from curative to palliative care in their activities.

In short, the higher the level of aggregation to which urban safety is pushed, the harder it becomes to distinguish its activities from those of peacebuilding. And, similarly, the lower the level of aggregation to which peacebuilding is pulled down, the harder it becomes to distinguish its activities from those of urban safety. The two come to merge together, sometimes indistinguishably.

The problem of aggregation is complemented by that of *politicization*. It is important to return to our introductory point that urban safety practices are *global* in scope. Because they are fundamentally related to problems caused by urbanization across the world – whether in Brussels, Beijing, or Bangui – and problems relating to cities whether they are urbanizing rapidly or not, they are necessary everywhere. Urban safety schemes are ongoing in London, just as they are in Kinshasa. One important consequence of this is that urban safety schemes are rarely *highly* politicized, not least because they are often part of a broader agenda focusing on sustainable urban development or the notion of 'smart cities' (see e.g. UNECE 2016) – now subsumed under Sustainable Development Goal 11 of the 2030 Agenda for Sustainable Development, which targets 'inclusive, safe, resilient and sustainable' cities and human settlements (UNGA 2015). Few people complain about road traffic safety measures on principle. Yet we tend to overlook this fact when exploring the relationship between urban safety and peacebuilding, not least because these two approaches are usually only found in combination in places outside rich and developed countries, where political and societal dynamics differ considerably.

Indeed, peacebuilding schemes are – as the name suggests – generally implemented following or during some degree of serious *political* violence and conflict, and/or extensive (and often organized) *criminal* violence, as found in parts of Central and Latin American (for a discussion see Krause 2009). Our focus when discussing the confluence of urban safety and peacebuilding practice is thus necessarily on states where this has been occurring, and where the governance context is intrinsically politicized. This frequently means that when we *do* look at self-declared urban safety schemes carried out alongside peacebuilding initiatives, the former are often more politicized than they would be under 'normal' circumstances. As a result, they may be involved in programming activities that look similar to a peacebuilding remit (e.g. engagement with community organizations to change 'ideas') rather than comprehensive urban planning schemes. But such politicization is *not* innate to urban safety.

Beyond sheer necessity (the persistence of crime, violence, and societal conflict in rapidly urbanizing settings), the confluence of urban safety and peacebuilding practice also occurs because of a mutually reinforcing logic. Indeed,

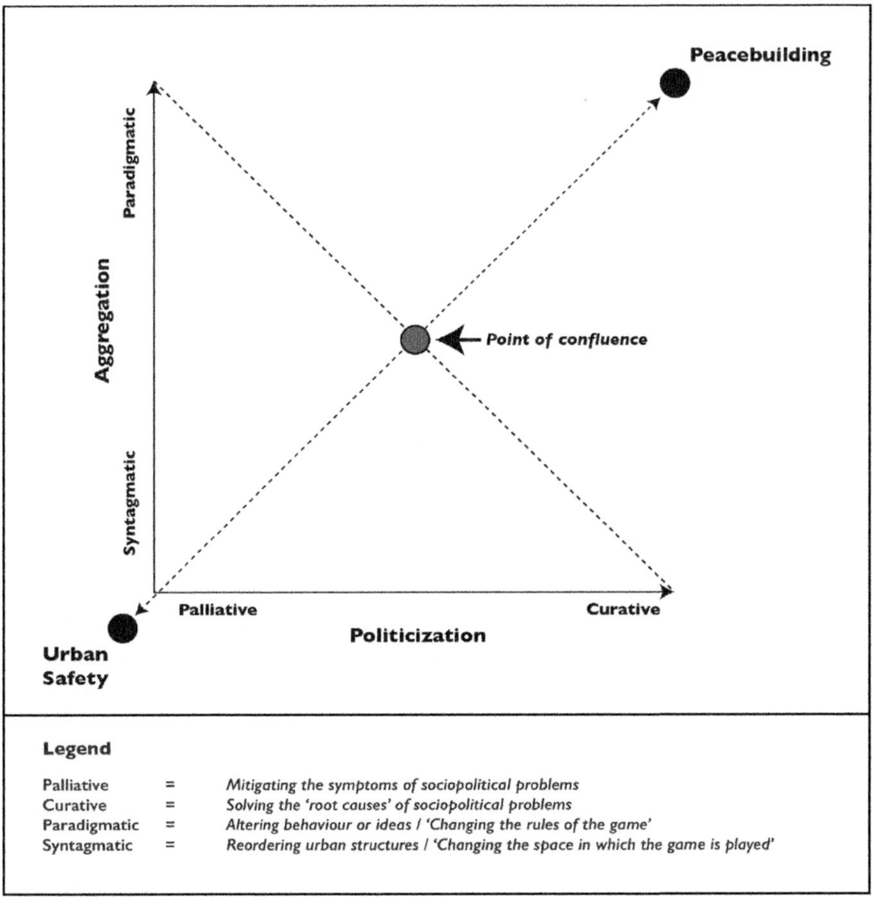

Figure 1.1 The confluence of urban safety and peacebuilding by aggregation and politicization.

while peacebuilding must be accepted as a political activity in and of itself, its success is premised on the *diminishment* of its own political status and – eventually – its complete disappearance. Yet the same cannot be said of urban safety measures, which succeed in so far as they remain apolitical and are not pulled into politics but maintain themselves as neutral background factors mitigating the perpetual hazards of any urban fabric. Hence in Table 1.1 we describe the temporality of urban safety as 'constant' and that of peacebuilding as 'linear' (or at the very least 'cyclical'). Peacebuilding seeks to take certain measures at time X to solve a conflict at time Y, at which point its activities will – ideally – no longer be required. Urban safety measures, by contrast, need to be *constantly* present and active in the urban fabric. There is a paradox here, however. Peacebuilding seeks to become *less* political by normalizing its programmes and ideas until they are 'accepted' and peace is – yes – 'built', but in doing so it is essentially trying to convert itself into an urban safety approach by moving down into carrying out more granular activities. In the process, however, it simultaneously works to politicize urban safety as an approach. These centripetal forces work alongside simple necessity to result in the paradoxical acceleration of the confluence of urban safety and peacebuilding programmes today.

That confluence is depicted schematically in Figure 1.1. It shows how we can chart the confluence between urban safety and peacebuilding along the two axes of aggregation (which correlates with the semiotic distinction between syntagmatic and paradigmatic forms) and politicization (which correlates with palliative or curative forms of care). The challenge now is to reflect on how this confluence between urban safety and peacebuilding practice, which has emerged largely by necessity, might be better formed into a truly 'integrated' approach.

Towards an integrated approach: dangers, challenges, and opportunities

Many of the contributions to this volume take the view that *sustaining* peace in urban environments requires identification of the contours of an *integrated* approach at the interface of urban safety and peacebuilding practice. Such integration is generally thought to be necessary because the 'confluence' that has emerged between the two approaches, as we have shown above, is largely a result of external forces, including a rise in the phenomena of concern to both urban planners and peacebuilders, a lack of wealth and/or material resources to carry out very granular urban safety practices in the global South, and a mutually reinforcing logic between urban safety and peacebuilding vis-à-vis their 'neutral' or 'political' status that drives them to convergence. While this 'shotgun wedding' of the two approaches has taught us much about their complementarity, we need to move towards more systematic integration of their respective strengths and weaknesses to avoid several dangers their thus far ad hoc combination; and we now conclude by identifying these.

The politicization of urban safety schemes seems likely to increase the fragmentation of city space into 'safe' gated communities and 'violent' peripheries (see e.g.

Landman 2012; Rodgers 2002; Caldeira 1996). That is, in situations of real or perceived threat or danger, the traditional focus of urban safety approaches on 'hazards' (a relatively neutral term) may be ignored, and its powerful means of intervening in the urban fabric of society utilized to divide cities through what Tunde Agbola (1997) has termed a new 'architecture of fear'. Indeed, take the example of the Israeli–Palestinian conflict again. The decades-long 'stall' in its peace process has seen peacebuilding efforts shift more and more towards traditional urban safety schemes on both Israeli and Palestinian sides. This has resulted in the manipulation of the urban fabric of the cities and their environs so as to 'separate' the two sides through punitive means (walls, checkpoints, etc.) that prevent economic development, political expression, and freedom of movement. The Israeli Committee against House Demolitions, which has campaigned against successive Israeli governments' manipulation of urban planning law to demolish and reorder urban space, refers to this as the construction of a 'matrix of control' able to direct minutely the lives of ordinary people through:

> … a maze of laws, military orders, planning procedures, limitations on movement, Kafkaesque bureaucracy, settlements and infrastructure – augmented by prolonged and ceaseless low-intensity warfare – that serves to perpetuate the Occupation, to administer it with a minimum of military presence and, ultimately, to conceal it behind massive Israeli 'facts on the ground' and a bland façade of 'proper administration'.
>
> (Halper 2009: 32)

In other words, the politicization of urban safety approaches to preventing crime, violence, and societal conflict may increase the 'bunkerization' of privileged societal groups at the expense of others. This can be seen in the creation of gated communities, the similar 'matrices of control' employed by military occupations in Baghdad, Palestine, southern Turkey, and elsewhere, and, in fact, in the more mundane urban safety schemes implemented in wealthy states. Take a trend towards 'hostile [or 'defensive'] architectures' that include the use of 'anti-homeless spikes' (studs placed on flat surfaces near doorways to stop homeless individuals sleeping there), slanting windowsills that prevent people sitting on them, and benches with arm rests to prevent people lying down (Petty 2016; Kinder 2014). These measures, as also analysed by López Rueda and Cara in this volume, all work to remove the 'right to the city' as it has been traditionally understood (Lefebvre 1996), and can be seen as a direct consequence of the politicization of urban safety through its integration of peacebuilding precepts in contexts where the 'root causes' approach of the latter has failed to gain adequate leverage over a 'fragile' urban environment. The result is a profoundly anti-political form of governance (as analysed by Ferguson 1990), something an integrated approach may want to avoid at all costs.

Alongside the challenges related to the politicization of urban safety are those caused by increasing aggregation. As we noted above, one of the reasons why it has been difficult to distinguish between urban safety and peacebuilding practice in contexts were these are both implemented relates to the fact that extensive and materially embedded urban safety approaches are costly and hence not frequently affordable for city municipalities. The result has been a shift to the meso level described in Table 1.1, where municipal governance becomes the entry point for interventions rather than the urban fabric in and of itself. By changing people's *ideas* about what it means to generate representative institutions of local *government* (as opposed to at times mere lip-service to vague notions of urban *governance*; see Jütersonke and Kartas 2015), it is hoped that the city will become a safer space. Paradoxically, however, this risks a focus on the 'duties' of individuals to the city as opposed to their 'right' to live in a safe and secure urban fabric. In other words, it occludes the socioeconomic rights of urban dwellers to a problematic extent, rendering them neoliberal subjects responsible for their own safety and that of others, without state intervention in the material fabric of the city.

This last point, however, should not be read as a critique of groups who advocate a focus on individuals. It must be seen, in fact, as a wider critique of peacebuilding and its apolitical stance vis-à-vis the necessity of wealth distribution to developing cities to improve their safety. Rather than attempting to reduce its political status, the socioeconomic aspect of urban safety is something peacebuilding schemes should perhaps be advocating more explicitly. However, in the present ad hoc combination of the two approaches, and in light of the politicization of peacebuilding in the context of shifting multilateral peace and security agendas as well as the concomitant interests of donors and host countries alike, this is not something that is likely to occur easily. With these dangers firmly in mind, then, how might we actually move towards an integrated approach? For opening directions, let us take Kim Dovey's words:

> One test for a good city lies in the capacity to walk the streets in safety – day or night, rich or poor, male or female, black or white, old or young. But another test lies in the capacity for all its citizens to gain access to the overwhelming vitality, creativity and diversity of urban life. The task is not to choose between, but rather to reconcile, these imperatives. It is one of understanding, managing and engaging with safety and danger in a creative and civilized manner. The struggle against the privatization and tranquilization of public space will be a long one. But the stakes for future generations are high and we will all be judged on the kind of city we bequeath them.
>
> (Dovey 2000: 13)

The task of an integrated approach to urban safety and peacebuilding requires, these words suggest, a better appreciation of the ideal-type roles described for them in this chapter. Urban safety is vital, in so far as its palliative care approach to the city ensures we can all 'walk the streets in safety' despite the continued presence of the hazards that mark any built environment. Those hazards – whether material or social – are, however, part of what constitutes the very 'creativity and diversity' of urban life, and the challenge is how to maintain and harness the positive energies associated with urban propinquity in a safe and inclusive manner. From this perspective, peacebuilding schemes should be seen as necessary in their rather more high-political role of dealing with the 'root causes' of particularly deleterious forms of crime, violence, and societal conflict. This role should ensure that peacebuilding engages the creative potential of urban politics but, in a truly integrated approach, does not allow that engagement to subsume the vital importance of acknowledging the materiality of the urban fabric of safety (and indeed encroach on the resources available to pursue activities related to it). In other words, an engagement with the politics of the city – with changing values, norms, and a willingness to fulfil one's duties to the city – must be integrated with an understanding that the very design and planning behind visions of urban safety are part of the fundamental right to the city that is being constantly negotiated as the urbanization of the planet continues. To assist in the formulation of a 'smart' and 'safe' city, we therefore require both curative care for its more serious ills and palliative care for the ailments that are – paradoxically – often the source of its vitality. To achieve this, researchers and practitioners should be careful to recognize the value of an integrated *division of labour* between urban safety and peacebuilding that respects the distinct foci of their ideal-type forms even where those ideal types cannot be full realized. An integrated approach – to end with semiotics – rests on us working towards crafting a better grammar for the urbanizing city.

References

Agbola, T. 1997. *Architecture of Fear: Urban Design and Construction Response to Urban Violence in Lagos, Nigeria*. Ibadan: African Book Publishers.

Austin J. 2016. 'Torture and the Material-Semiotics of Violence Across Borders', *International Political Sociology* 10(1), 3–21.

Barnett, M., H. Kim, M. O'Donnell, and L. Sitea. 2007. '"Peacebuilding": What Is in a Name?', *Global Governance* 13(1), 35–58.

Barthes, R. 2013. *The Language of Fashion*. London: Bloomsbury.

Bocco, R. and L. Takkenberg (eds). 2009. 'UNRWA and the Palestinian Refugees 60 Years Later', *Refugee Survey Quarterly* 28(2–3).

Cai, K. and J. Wang. 2009. 'Urban Design Based on Public Safety – Discussions on Safety-Based Urban Design', *Frontiers of Architecture and Civil Engineering in China* 3(2), 219–227.

Caldeira, T. P. R. 1996. 'Fortified Enclaves: The New Urban Segregation', *Public Culture* 8(2), 303–328.

Ceccato, V. 2012. 'Introduction', in Vania Ceccato (ed.) *The Urban Fabric of Crime and Fear*. London: Springer, 3–35.

Chetail, V. and O. Jütersonke (eds). 2014. *Peacebuilding: Critical Concepts in Political Science*. Abingdon and New York: Routledge.

Cure Violence. 2016. 'The Cure Violence Public Health Model', available at http://cureviolence.org/the-model/essential-elements/ (accessed 2 July 2018).

Dana, T. 2013. 'Palestinian Civil Society: What Went Wrong?', *Al-Shabaka*, 14 April, available at https://al-shabaka.org/briefs/palestinian-civil-society-what-went-wrong/ (accessed 2 July 2018).

Debiel, T., T. Held, and U. Schneckener (eds). 2016. *Peacebuilding in Crisis: Rethinking Paradigms and Practices of Transnational Cooperation*. London: Routledge.

Dovey, K. 2000. 'Redistributing Danger: Enclosure and Encounter in Urban Design', *Australian Planner* 37(1), 10–13.

Ferguson, J. 1990. *The Anti-Politics Machine: 'Development', Depoliticization, and Bureaucratic Power in Lesotho*. Cambridge: Cambridge University Press.

Galtung, J. 1976. 'Three Approaches to Peace: Peacekeeping, Peacemaking, and Peacebuilding', in J. Galtung, *Peace, War, and Defense: Essays in Peace Research*, Vol. 2. Copenhagen: Christian Ejlers, 282–304.

Galtung, J. 1996. *Peace by Peaceful Means*. Oslo: Peace Research Institute Oslo.

Geneva Peacebuilding Platform. 2015a. 'Concept Note: Technical Working Group on the Confluence of Urban Safety and Peacebuilding Practice.' Geneva: Geneva Peacebuilding Platform.

Geneva Peacebuilding Platform. 2015b. 'White Paper on Peacebuilding.' Geneva: Geneva Peacebuilding Platform.

Halper, J. 2009. *Obstacles to Peace*. Jerusalem: Israeli Committee against House Demolitions.

Heathershaw, J. 2008. 'Unpacking the Liberal Peace: The Dividing and Merging of Peacebuilding Discourses', *Millennium* 36(3), 597–621.

Iwami, T. 2016. 'Understanding Japan's Peacebuilding in Concept and Practice', *East Asia* 33(2), 1–22.

Jütersonke, O. and M. Kartas. 2015. 'The State as Urban Myth: Governance Without Government in the Global South', in R. Schuett and P. M. Stirk (eds) *The Concept of the State in International Relations: Philosophy, Sovereignty, Cosmopolitanism*. Edinburgh: Edinburgh University Press, 108–134.

Jütersonke, O. and K. Krause. 2013. 'Peacebuilding and the City: Setting the Scene', Geneva Peacebuilding Platform Brief 9. Geneva: Geneva Peacebuilding Platform and Centre on Conflict, Development and Peacebuilding.

Jütersonke O., K. Krause, and R. Muggah. 2007. 'Guns in the City: Urban Landscapes of Armed Violence', in *Small Arms Survey 2007: Guns and the City*. Cambridge: Cambridge University Press, 161–195.

Kinder, K. 2014. 'Guerrilla-Style Defensive Architecture in Detroit: A Self-Provisioned Security Strategy in a Neoliberal Space of Disinvestment', *International Journal of Urban and Regional Research* 38(5), 1767–1784.

Kostof, S. 1991. *The City Shaped: Urban Patterns and Meanings through History*. London: Thames and Hudson.

Krause, K. 2009. 'Beyond Definition: Violence in a Global Perspective', *Global Crime* 10(4), 337–355.

Landman, K. 2012. 'Reconsidering Crime and Urban Fortification in South Africa', in V. Ceccato (ed.) *The Urban Fabric of Crime and Fear*. London: Springer, 239–264.

Lefebvre, H. 1996. 'The Right to the City', in H. Lefebvre, *Writings on Cities*. Oxford: Blackwell, 147–159.

Lei, Z. 2011. 'Two Pillars of China's Global Peace Engagement Strategy: UN Peacekeeping and International Peacebuilding', *International Peacekeeping* 18(3), 344–362.

Levy, J. M. 2016. *Contemporary Urban Planning*. London: Routledge.

Paris, R. 2002. 'International Peacebuilding and the "Mission Civilisatrice"', *Review of International Studies* 28(4), 637–656.

Petty, J. 2016. 'The London Spikes Controversy: Homelessness, Urban Securitisation, and the Question of Hostile Architecture', *International Journal for Crime, Justice, and Social Democracy* 5(1), 67–81.

Richmond, O. P. 2006. 'The Problem of Peace: Understanding the "Liberal Peace" ', *Conflict, Security and Development* 6(3), 291–314.

Rodgers, D. 2002. *'We Live in a State of Siege': Violence, Crime, and Gangs in Post-Conflict Urban Nicaragua*. London: London School of Economics and Political Science.

Sebok, T. A. 2001. *Signs: An Introduction to Semiotics*. Toronto, ON: University of Toronto Press.

Sivak, M. and O. Tsimhoni. 2008. 'Improving Traffic Safety: Conceptual Considerations for Successful Action', *Journal of Safety Research* 39(5), 453–457.

Stepanova, E. 2004. 'War and Peace Building', *Washington Quarterly* 27(4), 127–136.

UNECE (UN Economic Commission for Europe). 2016. 'Shaping Smarter and More Sustainable Cities: UNECE and ITU Launch the United for Smart Sustainable Cities Global Initiative', press release, 19 May. Geneva: UNECE.

UNGA (UN General Assembly). 2015. 'Transforming Our World: The 2030 Agenda for Sustainable Development', UN Doc. A/RES/70/1, 21 October. New York: United Nations.

UN-Habitat (UN Human Settlements Programme). 2007. *Global Report on Human Settlements: Enhancing Urban Safety and Security*. London: UN- Habitat/Earthscan.

UN-Habitat (UN Human Settlements Programme). 2012. 'Safer Cities Programme', available at http://unhabitat.org/urban-initiatives/initiatives-programmes/safer-cities/ (accessed 2 July 2018).

UNSC (UN Security Council). 2016. 'Resolution 2282: Post-Conflict Peacebuilding', UN Doc. S/RES/2282, 27 April. New York: United Nations.

UNSG (UN Secretary-General). 1992. 'An Agenda for Peace', UN Doc. A/47/277, 27 June. New York: United Nations.

Wennmann, A. 2010. 'Grasping the Strengths of Fragile States: Aid Effectiveness Between "Top-Down" and "Bottom-Up" Statebuilding', CCDP Working Paper 6. Geneva: Centre on Conflict, Development and Peacebuilding.

2 Urban safety and crime prevention

Architectural perspectives from Quito and Guayaquil

Cyntia López Rueda and Luna Cara

Introduction

The spatial configuration of the public realm represents the urban environment in which the life of citizens takes place. It also shapes the social fabric of a city and influences how citizens behave in private and public spheres. How this space is configured has also an important bearing for how citizens perceive their safety within the urban space (Escudero López 2016: 2).

Building on the distinction between safety and security, and on the observation that this distinction is blurred in Spanish through the common terminology of *seguridad*, this chapter explores the relationships between architectural typologies and the spatial configuration of cities, understood as a (culturally and linguistically shaped) social product (Lefebvre 2007 [1974]: 35). The study of two public spaces in Quito and Guayaquil (Ecuador) reveals the existence of different design strategies related to crime prevention and urban safety. The chapter argues that people's perception of urban safety is influenced by the spatial connection existing between the shape of public places and the set of architectural typologies surrounding them.

The chapter focuses on the historic centres of Quito and Guayaquil – specifically Quito's 24 de Mayo Boulevard and Guayaquil's Simón Bolívar Waterfront. While both public spaces were planned as part of urban interventions at the neighbourhood level in an effort to regenerate degraded areas, the two spaces are characterized by different architectural typologies in their private cores. Contrasting design intentions can be discerned through this juxtaposition of public and private, and analysed through the lens of architectural and urban design strategies related to safety and crime prevention.

Conceptualizations of safety in urban and architectural design

In his work *The Production of Space*, Henri Lefebvre (2007 [1974]: 53) argues that space is the result of a set of social relations and practices that cannot be replicated, since they are a unique social product that is continuously being transformed. Indeed, the city space endures a myriad of transformations, among which the conceptualization and design practices related to safety and security stand out.

In Spanish, *seguridad urbana* has been described as 'the situation of public tranquillity and free exercise of individual rights, whose effective protection is entrusted to the forces of public order' (Real Academia Española, 2014). Of note here is the emphasis on public tranquillity and particularly the protection of citizens, which has arguably been a hallmark of contemporary urban public policies in Latin America. As a result, debates have been dominated by terms such as 'crime prevention' and 'violence reduction' – terms that are also part of the social imaginary when engaging with the spatial *problèmatique* of urban safety. The terminology of urban safety, in turn, is boiled down to actions conceived to decrease the deep fear of crime and the critical need for protection of the inhabitants (Cara and López Rueda 2017: 98).

Anglophone debates – e.g. Carmona *et al.* (2010), Minton (2012), Raco (2007), Glaeser (2012), and Dovey (2000), among others – point to a semantic difference between urban safety and crime prevention. It is a differentiation that corresponds to the ideology of the spatial configuration and design of public and private places. Indeed, the production of safe cities has not only become a major topic of global concern but is also a critical factor for the strategic planning of sustainable communities. In consequence, several academic studies have emphasized the minimum factors that should be considered when designing for public safety. It is possible to distinguish between four public safety factors for the design of urban space: psychological safety, defence for safety, behavioural safety, and safety against disaster (Cai and Wang 2009: 221), as depicted in Figure 2.1.

For the present purposes, the focus is solely on the first two factors: psychological safety and defence for safety. Moreover, terrorist attacks, which Cai and Wang also analyse, will not be addressed in the cases of Quito and Guayaquil. *Behavioural safety* deals with damage from environmental disturbance when people are conducting common activities (Cai and Wang 2009: 221), but the functioning of this component is considered as an outcome of the configuration of the space studied under the guise of the first two factors. Similarly, this research does not focus on *safety against disaster*, as the public spaces analysed do not face natural disasters regularly and are thus not designed accordingly.

Psychological safety

From the perspective of urban design, the issue of human needs has been discussed by a number of academic studies, notably by Abraham Maslow (1987), who covers concerns of both private and public behaviour. As reported by Maslow, once the basic physiological needs (survival) are partially satisfied, human beings will search for more abstract and aesthetical needs related to the psychological pursuit of safety (Carmona *et al.* 2010: 134; Lang 2005b: 13). The environmental stress theory states that cognitive factors (those acquired

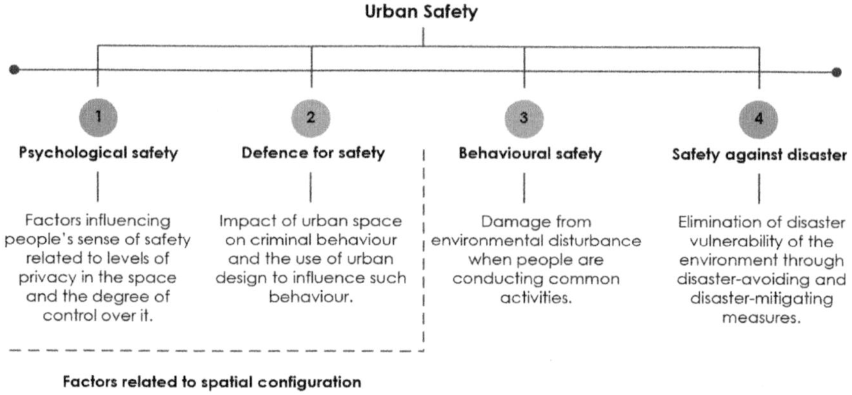

Figure 2.1 Safety factors in the design of urban space.

through experience and the senses) and autonomic factors (those controlled by the nervous system) scan the environment to evaluate the stimuli there as threatening or not (Lazarus and Cohen 1977: 94). That is to say, people assess the nature of a space through cognition of environmental stimuli and then evaluate the environment as safe or unsafe through the cognitive process (Cai and Wang 2009).

Most of the contemporary literature and academic studies regarding urban violence and safety concerns show that the perception of safety or danger does not always relate directly to actual incidence of crime. Regarding specific spatial factors, a sense of safety is mainly related to people's privacy level in and degree of control over environmental factors. Lang (2005a) states that the privacy level is directly associated with the protection scale that people sense in the space, while the control degree is connected to people's ability to avoid or amend factors of risk (cited in Cai and Wang 2009: 221).

In addition to privacy and control, Lynch (1960: 2–5) argues that 'legibility' is a factor that increases people's capability of spatial orientation, and therefore reduces environmental fear and psychological anxiety. The legibility of the cityscape is the ease with which a person can organize and recognize the elements of a particular place in a coherent image. Moreover, in her pioneering work Jane Jacobs (1961) argues that psychological safety does not require external vigilance but that citizens' surveillance is crucial. Jacobs stressed the need for urban activity to provide citizen surveillance and that a successful area was one in which 'a person must feel personally safe and secure on the street among all these strangers' (ibid.: 30). Jacobs argues for *eyes on the street* – that is, surveillance for the natural proprietors of the street, where buildings are equipped to handle strangers by their orientation to the street and to the public places (ibid.: 35).

In brief, psychological safety is linked to the spatial configuration of urban life. Consequently, the spatial strategies that foster legibility are related to privacy level and degree of control of the place.

a *Linearity.* Linear areas are those where local spatial conditions present a continuous constitution on both sides of the roads and provide a legible image and in general a perception of psychological safety (Hillier and Hanson 2004: 36). For instance, the presence of continuously lined dwelling entrances or the same building height facing a public space from both sides of the road contribute to facilitating a sense of orientation in a particular area.

b *Number of secondary access points.* To contribute to a sense of orientation and safety, the buildings in a neighbourhood should avoid the presence of secondary access points. Citizens are most likely to sense vulnerability to environmental factors when buildings have a secondary access in side or back alleys (ibid.: 37).

c *Synergy* is related to how local movement potentials *within* an area relate to movement potentials *through* the area – i.e. the relationship between the local and the global structures of a particular place (ibid.: 42). The zonal planning of the modern city separates global circulations from local routes to accelerate the flow of vehicle traffic, which isolates areas and prevents citizens from perceiving the overall context of a place.

d *Connectivity* refers to the directness of links and the density of connections in a street network (Lynch 1960). For instance, the presence of minimal dead ends, more intersections in an area, and small blocks (75–150 metres) contributes to a sense of connectivity, since they increase vitality and walkability (Jacobs 1961: 178–186). The analysis here focuses on the number of exits or 'ways out' of a particular street, using a range of 1 (low connectivity) to 6 (high) in terms of coding street types. Roads that present only one way out are coded as a Level 1 (such as dead-end streets), while a fully connected street section where a pedestrian could take different directions are coded as a Level 6 street.

e *Visual connection* describes the quantity of visual impediments found in an area. Visibility in a certain space fosters legibility and a sense of safety, since inhabitants and strangers could control the area from their locations regardless of whether they are public or private (see Figure 2.2).

Defence for safety

Using spatial design to impact citizens' behaviour – and in particular on issues related to security, crime, and urban violence – has a long tradition (Cozens 2008: 153). The theory of crime prevention through environmental design (CPTED), later known as 'designing out crime', states that 'the proper design and effective use of the built environment can lead to a reduction in the fear and incidence of crime, and an improvement in the quality of life' (Crowe and Fennelly 2013: 280). This means it is preferable to identify the conditions of the physical environment that provide opportunities for criminality, and then modify these spatial configurations to reduce and prevent crime.

The Death and Life of Great American Cities (Jacobs 1961) was the first influential work suggesting that the existence of active street life and certain characteristics of the built environment could reduce the opportunities for crime. In the early 1970s the prevention of delinquency and crime focused on the manipulation of the built environment. Fundamental to these theories is the work of Oscar Newman (1972) and his concept of 'defensible space' (see Geason and Wilson 1989: 3).

Newman's ideas on establishing a link between the quality of architectural design and crime rates have had a profound influence on both urban and architectural designers over the last 40 years. The author associates the organization of architectural and urban space with inhabitants' behaviours, arguing that it is possible to create a sense of territoriality that allows natural surveillance of the exterior dynamics. Following Newman's theory, several authors in academia and public policy have outlined spatial strategies to construct the imagery of the aforementioned territoriality to promote residents' natural surveillance.

The present research is focused on selecting the CPTED strategies that are directly related to the predominant typological models of the neighbourhoods in which the case studies are located, to establish a coherent connection between the private and the public way of living in and producing space. The strategies that have been considered are the following.

a *The accuracy of the spatial hierarchy.* To simplify natural surveillance, the spatial structure of the city must allow direct recognition of territorial divisions in relation to use: public, semi-public, private, and semi-private zones are easily recognizable. In Newman's work (1996: 15), semi-public spaces are public areas that have a set of common and universally acceptable rules regarding behaviour and access. By contrast, semi-private places have a higher level of privacy, since the rules regarding behaviour and access are set by a limited number of inhabitants. The *definition of accurate boundaries* between spatial hierarchies avoids the formation of ambiguous spaces where, according to CPTED theory, the crime rate is expected to be higher due to the lack of territoriality. The transit between private and public spaces should be progressive, with transitional spaces that change physical characteristics or present visible limits, among other design strategies (Geason and Wilson 1989: 16). The accuracy of spatial hierarchy is related to the concept of placelessness, where 'place' is explained as a positive sense of identification with spaces that inhabitants appreciate, while placelessness is the absence of such identification (Dovey 2016: 112).

b *Building height and façade composition.* Contact with ground-level activities and events is only possible from the first few floors in a multistorey building; that is to say, the ability to have contact with the ground level decreases from the third and fourth floors up (Gehl 2011: 98). Buildings that are directly linked to public spaces should allow natural supervision and are

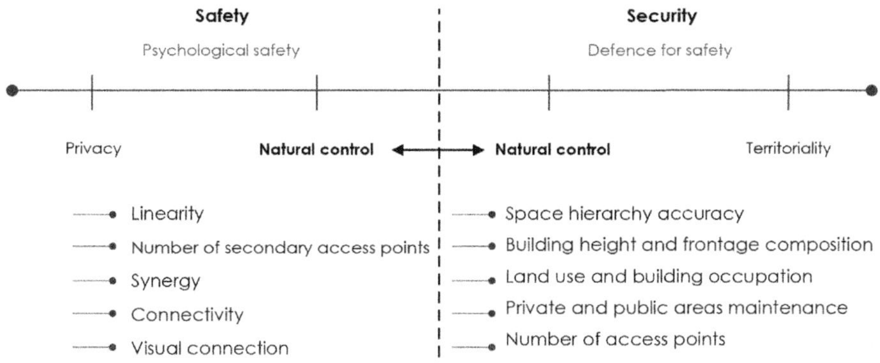

Figure 2.2 Spatial strategies for psychological safety and defence for safety.

not to be higher than three stories, to control the external areas visually and, accordingly, permit natural supervision.

c *Land use and building occupation.* To promote natural policing in a place, it is important to acknowledge the activities of citizens in the neighbourhood. Surveillance at night occurs naturally in residential or mixed-use zones. In contrast, exclusively retail zones combined with unused buildings increase the possibility of crime, particularly in the evening hours.

d *Private and public area maintenance.* One of the most important premises of CPTED theories is that the security of a space is directly linked to the inhabitants' perceptions of the place, as urban decay is more likely to attract criminal acts (Geason and Wilson 1989: 18; Palimariciuc 2016: 15). The maintenance of public and private spaces determines the aesthetic quality of a place, and, according to CPTED theories, the general image of the frontages, public spaces, and street furniture has a bearing on criminal activity. As the 'broken windows' theory of crime prevention argues, when a window in a building is broken and not repaired, the rest of the windows will soon follow suit (Carmona *et al.*, 2010: 328).

e *Number of access points.* The literature on 'designing out crime' states that crime rates are increased when there are multiple access points to a single residential complex, as it is then easier to leave a certain space (Geason and Wilson 1989: 35). In addition, CPTED states that access points must be easily identified and have physical barriers to discourage strangers from entering the private complex (see Figure 2.2).

The relationship between psychological safety and defence for safety: safety and security

It is now widely accepted that there are two main approaches to engaging in urban safety issues, involving either 'dispositional' or 'situational' measures

(Clarke 1980: 136; Carmona *et al.* 2010: 150). Criminological theories identify the dispositional approach as focusing on the motivations to commit criminal acts and thus on reducing the individual's *disposition* towards them through education and moral guidance. On the other hand, *situational* measures exploit the social, psychological, and physical settings to reduce the opportunity for criminal behaviour (Clarke 1997: 2). Dispositional measures are not related to the design of the formal configuration of space, hence the spatial analysis outlined in this chapter is based solely on situational considerations.

Intuitively, there would seem to be a strong bond between spatial configuration and users' behaviour in public places; however, the degree to which urban form determines behaviour is unclear empirically. Determinism is the idea that the environment can determine or cause certain kinds of behaviour: if the built environment is changed, behaviour will also be improved (Dovey 2016: 41). This line of thought implies that social problems could be solved by physical design, a statement that the contemporary urban literature has refuted. Indeed, the word 'determinist' is often used to communicate a false belief.

According to Dovey, the physical environment cannot determine behaviour but it is not neutral either, since built form mediates social practices through a combination of enabling and constraining. 'At the most banal level a wall constrains movement and enables privacy, but it does not cause any kind of behaviour' (Dovey 2000: 10). The human environment can be seen in terms of *overdetermination* and *underdetermination*, where overdetermination in design locks the usage of a place into a certain way. While design cannot determine that a space will be used in the 'right' way, it can determine that not much else will happen. By contrast, an underdetermined design serves a multiplicity of possible functions and creates a city where users are able to choose from multiple options of use (Dovey 2016: 42). At any rate, the analysis pursued here is based on the idea that the spatial configuration of built form cannot determine the evolution of a social problem as complex as crime, although it can influence the overall conception of urban safety.

Despite the high degree of similarity between spatial strategies geared towards a sense of psychological safety and those targeting security, there are different ways in which urban form tackles safety. Hillier and Hanson (2004: 31) recognize two divergent views that are outlined in the work of Jacobs and Newman: while Jacobs advocates open and permeable environments for both strangers and inhabitants, Newman explores closed and impermeable spaces where inhabitants are able to recognize strangers as intruders. Both of these urban design trends are considered as *situational* measures, focusing the attention on the prospects of crime occurring (Carmona *et al.* 2010: 151).

Nevertheless, there is a degree of overlap between Jacobs and Newman: namely the passive surveillance that Hillier (1996: 141) called 'continuous co-presence'. Co-presence is a measure of vitality linked to the existence of more than two people in the same visual field at any one time (Dovey 2016: 46). It can be said that for Jacobs (1961) the presence of strangers is a source of safety and humanizes the streets, while for Newman (1972) the presence of strangers is a clear source of danger and would be abnormal for inhabitants. Both of these conceptualizations deal with the search for privacy, territoriality, and sense of control over an area, but there are behavioural implications that legitimate different spatial circumstances in each case. The spatial models that have been recognized in the aforementioned situational measures are studied as urban 'encounter' and 'enclosure' models by Dovey (2000: 10), who emphasizes that the difference lies in the encouragement of or the reliance on an encounter with strangers. The encounter model theoretically developed by Jacobs (1961) promotes a highly permeable urban structure with short blocks and multiple connections in which contact with strangers is protective. Hillier and Hanson (1984: 18) describe this model as a space where 'strangers police the space, while the inhabitants police the strangers'.

In contrast, the enclosure model is perceived as a form of social control, since the production of urban safety moves seamlessly into the production of social sameness (Dovey 2000: 12). Walls, fences, gates, and other physical means for achieving defence for safety are meant to put distance between inhabitants and what they perceive as a threat. The closed public space gains urban recognition from the deterioration of the open public realm surrounding it.

Defensible space and CPTED theories declare that the way to change behaviour is by controlling the environment rather than by improving social conditions; by creating territoriality, the spaces would defend themselves by marking out clear boundaries (Minton 2012: 71). Spaces that emphasize the defence for safety factor are often empty areas that promote fear – instead of the safety that is achieved in places where people are free to circulate without physical boundaries (ibid.: 72). This design approach is consistently followed by public policies in spite of Jacobs's (1961) study, which is continuously praised as classic urban literature because of her emphasis on diversity in cities. In conclusion, the design of public space is left in a strange position of having police officers, rather than architects and urban designers, responsible for spatial configuration.

All in all, and as outlined in Figure 2.3, it can be said that the enclosure model is an explicit measure of social control that evolves naturally towards the gated

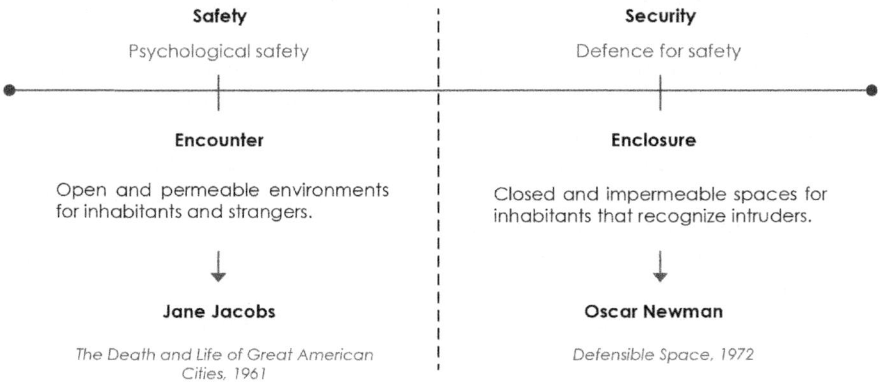

Figure 2.3 Psychological safety and defence for safety compared.

community, while the encounter model can be considered the result of a city with mixed functions and social diversity.

It is important to mention that the selection of the strategies dealing with psychological safety and defence for safety may seem overly simplified. However, it is essential that minimalist approaches are recognized, to clarify the basis of the concepts analysed in this chapter. Since both enclosure and encounter models are means of controlling urban safety, and therefore crime, it is crucial to understand that both models work together much of the time in specific neighbourhoods.

The analytical utility of these minimalist approaches is evidenced when designing public space, since these two currents of thought control and regulate the presence of strangers in public premises. Moreover, when analysing the strategies one can see that the enclosure model tends to eliminate Jacobs's eyes on the street (Jacobs 1961: 35), which should be a conscious design strategy rather than an unintended outcome. The strategies of both models analysed in this chapter are useful for urban designers, since they enable the study of the spatial characteristics at the macro level and allow us to deduce the possible behaviour of users in the surrounding streets.

Quito and Guayaquil: the relation between architectural typology and urban safety

According to Martí Arís (2014: 25), an architectural typology acts as a set of formal characteristics describing a spatial structure that remains unalterable over time and adapts to all kinds of circumstances. The concept of architectural type is understood as a formal pattern capable of multiple variations, i.e. it is not considered as a mechanism of systematic reproduction.

The process of typological classification completed in the present research is based on the theoretical contribution of architects of the postmodernist trend, such as Aldo Rossi (1984) and Carlos Martí Arís (2014). It is important to highlight that the typology of the buildings is analysed to expose the spatial patterns influencing the architectural strategies used in terms of urban safety and crime prevention.

In Quito the research focus was on the regeneration of 24 de Mayo Boulevard, located on the avenue of the same name. Five urban blocks with 65 buildings surrounding the public space were studied in terms of the spatial approaches taken with regard to urban safety and crime prevention (see Figure 2.4). Most of the edifices analysed are walk-up buildings of approximately 700 square metres on average. In Guayaquil the area analysed comprises the first section of Simón Bolívar Waterfront, located between 10 de Agosto Avenue and Francisco Paula de Icaza Street. Six urban blocks were examined, with 22 buildings directly related to the public space, and again applying spatial urban safety and crime prevention strategies (see Figure 2.5).

The spatial configuration of Guayaquil is directly related to urban regeneration policies that since 1992 have focused on issues related to security in and of

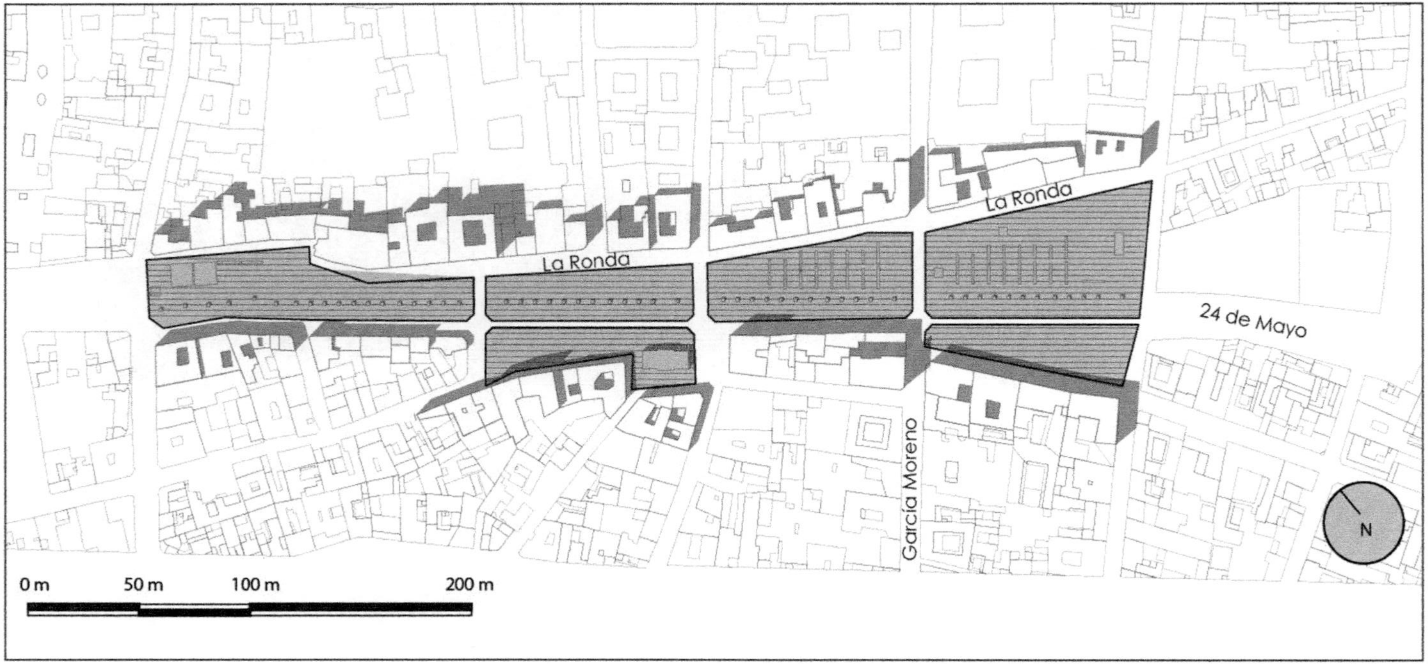

Figure 2.4 24 de Mayo Boulevard, Quito – site plan.

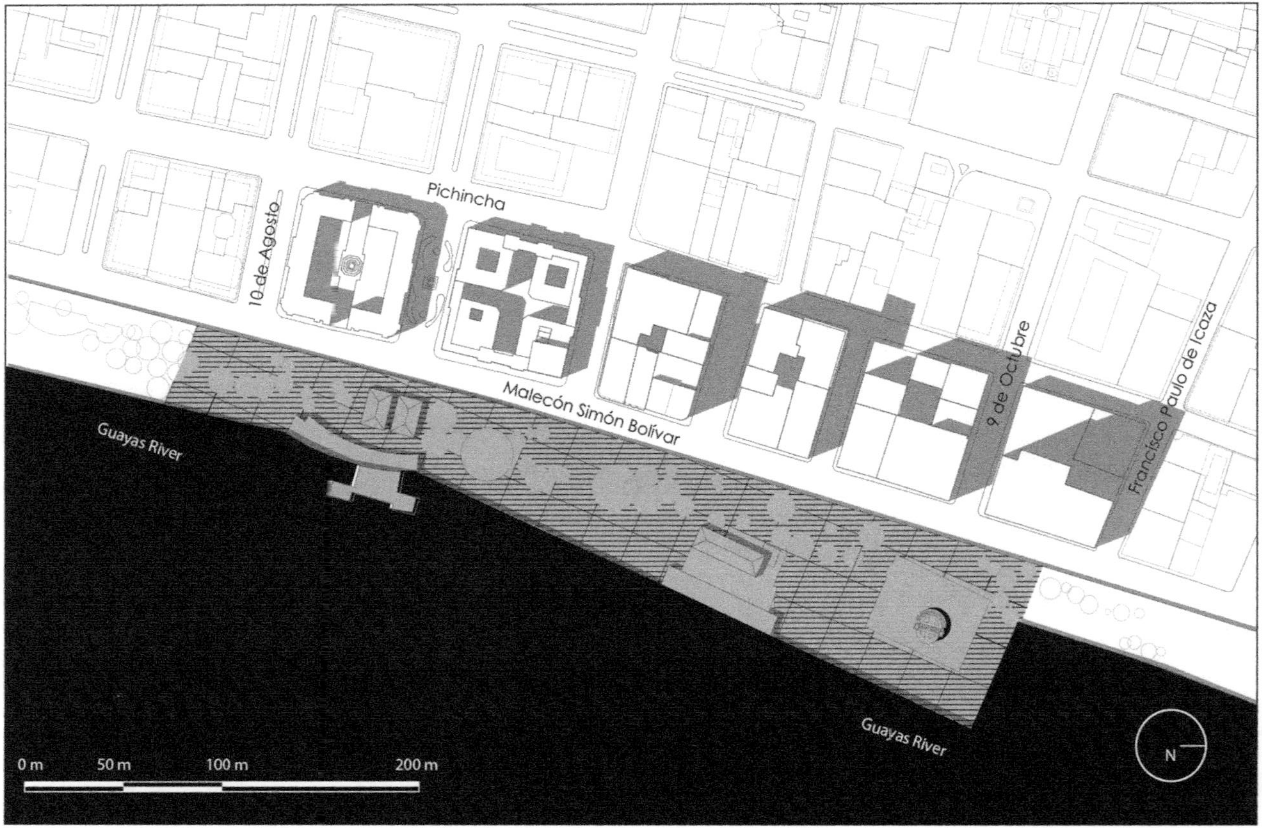

Figure 2.5 Simón Bolívar Waterfront, Guayaquil – site plan.

public spaces. During the two consecutive mayoral tenures of León Febres Cordero (1992–1996 and 1996–2000), Guayaquil saw a drastic spatial reconfiguration of its main public spaces (Wong 2005: 183). The implemented management model focused on road infrastructure, urban connectivity, and the regeneration of public areas with a public–private partnership (ibid.: 182). In addition, the regeneration of Simón Bolívar Waterfront (known as 'Malecón 2000') was considered a flagship project that was finally realized in 2000 under Jaime Nebot's mayoral tenure; Nebot hails from the same political party as Febres Cordero, representing Ecuador's conservative political spectrum.

In 2003 Guayaquil was still considered a violent city, with an insecurity index perception of 66.3 per cent – a figure that was partly based on real events and partly on other factors such as the weakness of the social fabric and the (mis)information provided by the media (Chiriboga 2006: 3). When the city's sea port was built in 1963, Guayaquil lost its connection with the river. Over subsequent decades the waterfront area, now suffering from disuse, became the symbol of the city's urban decay (Wong 2005: 184).

During the urban rehabilitation since 2003, security perceptions of inhabitants improved due to new investments that stimulated the homogenization of public spaces and increased public and private surveillance of commercial and tourist areas to move 'antisocial' activities to the outskirts of the city (Garcés 2004: 56). Guayaquil's urban rehabilitation has been acknowledged to have led to an increase in urban security, the banning of informal retail activities on the streets, and physical enclosures of the public spaces in the historic centre, such as gates and surrounding fences (ibid.: 60). One of these currently gated public spaces is Simón Bolívar Waterfront, where the spatial characteristics were drastically modified due to private control and surveillance.

By contrast, the regeneration of 24 de Mayo Boulevard in Quito was pursued through a public–private partnership that aimed to improve neighbourhood vitality through urban safety initiatives. The restoration and conservation of buildings with heritage value and the increase in student housing around the area were considered as crucial strategies to restore vitality and improve urban safety in the historic centre (Ortega 2014: 23).

This research contrasts the opposite approaches of producing 'safe' public spaces in Quito and Guayaquil to recognize design strategies for urban safety and crime prevention in public spaces.

Quito's historic centre

The predominant architectural typology identified in Quito's historic centre is the central courtyard, a feature introduced to Quito by the Spanish during the course of the sixteenth century (Silva 2001: 876). Throughout history, the courtyard archetype has evolved simultaneously with the architecture itself, since it is considered central to private living. The centralized organization around the courtyard regulates the areas surrounding it and represents the most intimate core of society (Micara 1985: 23).

In addition, the central courtyard is characterized by an interior void that creates an intangible limit between the interior and the exterior world. It is important to mention that the entrance is not directly related to the central void, with a transitional space called *zaguán*, or hallway, regulating the direct contact with the exterior (Silva 2001: 879). Furthermore, Quito's central courtyard typology does not entirely separate interior from exterior areas, since the spatial configuration considers voids at the façade of the buildings on the streets. The windows face both the central courtyard and the exterior, which allows natural surveillance by residents.

Finally, at the neighbourhood scale, when the model is analysed the central courtyard is understood as a space organizer, since it is conceived as a module that is repeated in space, generating not only the formal but also the social structure of the city. In summary, the location of the central courtyard module rules the relations between private and public spaces in Quito's historic centre (see Figure 2.6).

In terms of psychological safety, and using the spatial (CPTED) strategies outlined above, the analysis of 24 de Mayo Boulevard can be distilled as follows.

a *Linearity.* From the 65 buildings analysed, the study shows that 67 per cent present a continuous and homogeneous constitution on both sides of the road. If one considers that the boulevard is divided into four urban blocks along the avenue, three of these blocks present the linear constitution mentioned above, hence the sense of psychological safety in the boulevard is high due to the legibility of the place. The block in which the linear constitution of the space is lost is where public facilities are found on one side of the road, while on the other (east) side there are residential mixed-use buildings.

b *Number of secondary access points.* While around a quarter of all the buildings analysed have secondary access points, psychological safety is not affected due to the location of the entrances facing the boulevard or La Ronda Street, contributing to an active street life.

c *Synergy.* The analysis reveals that the morphology of the boulevard evidences the global structure of the whole neighbourhood: the public strip is an

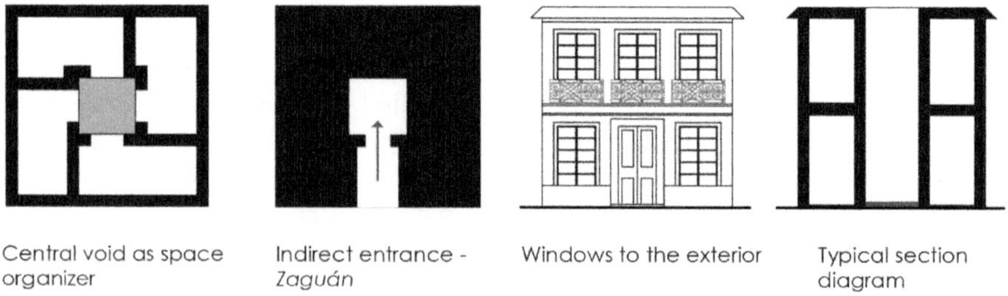

Figure 2.6 Quito: typological abstraction.

extension of the avenue, and therefore does not alter the perception of the whole urban component. In addition, the movement potentials in the neighbourhood do not isolate urban blocks or the public space studied, since every street around the space is a one- or two-lane road that permits a relationship between 24 de Mayo Boulevard and its surroundings (see Figure 2.7).

d *Connectivity.* As depicted in Figure 2.8, the area has a medium degree of connectivity: 25.9 per cent of the street sections analysed present low levels of connectivity, between levels 1 and 3 according to the diagram. On the other hand, 13 per cent of the segments display level 4 connectivity, 26 per cent level 5, and more than 48 per cent show the maximum level of connectivity, which contributes greatly to inhabitants' and users' psychological safety.

e *Visual connection.* The study shows minimal visual impediments preventing inhabitants from controlling the public space. Similarly, users are able to dominate the whole area visually from the boulevard. Due to its spatial configuration, 24 de Mayo Boulevard is presented as a legible space that fosters psychological safety (see Figure 2.9).

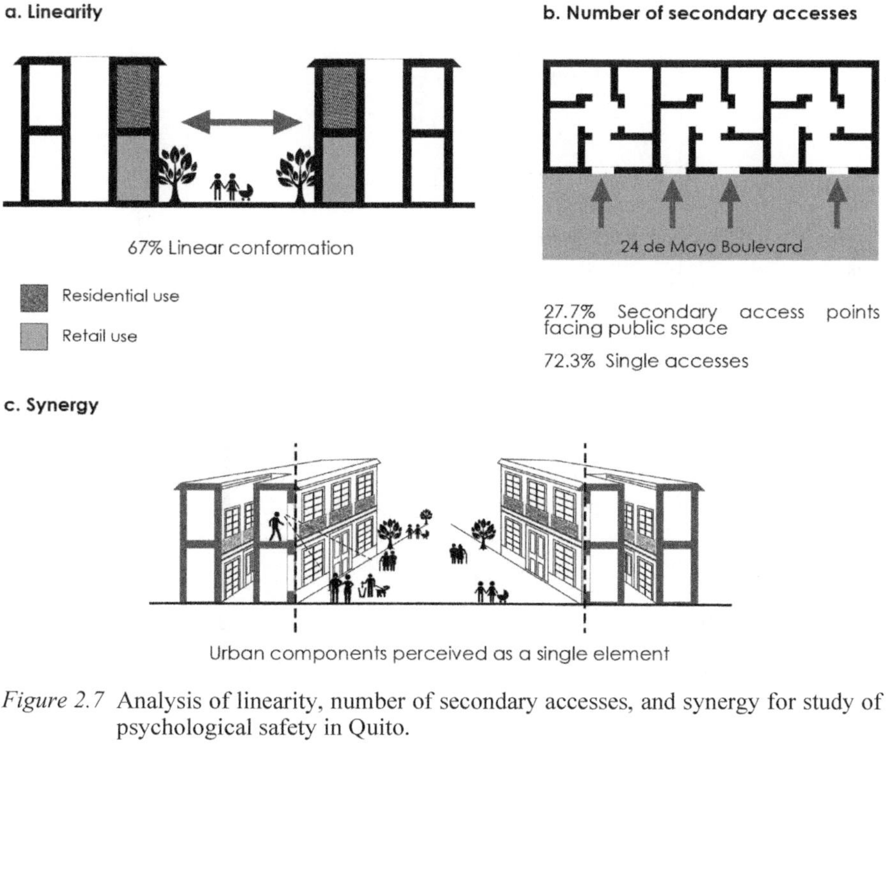

Figure 2.7 Analysis of linearity, number of secondary accesses, and synergy for study of psychological safety in Quito.

d. Connectivity

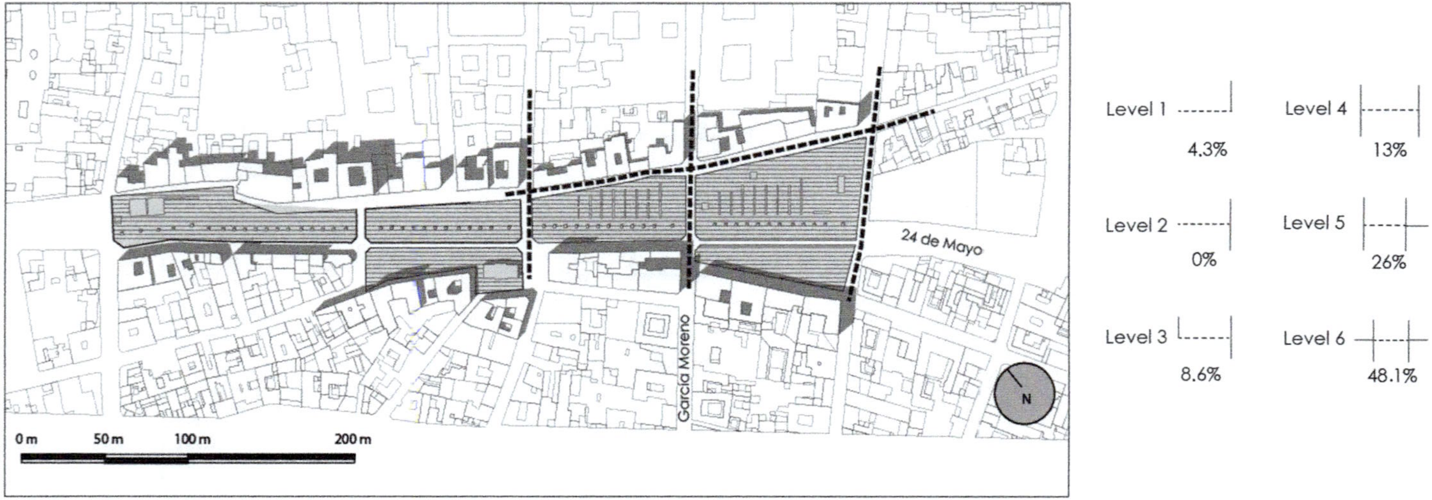

Figure 2.8 Connectivity study of 24 de Mayo Boulevard.

Figure 2.9 Visual impediments on the boulevard.

In terms of defence for safety, an analysis of spatial strategies offered the following insights.

a *The accuracy of spatial hierarchy.* The majority (70.77 per cent) of the buildings analysed have semi-public use, while the rest are completely private on the ground floor. On the other hand, 19.56 per cent of the upper floors are semi-public while the remaining percentage is for private use. In addition, 60 per cent of the buildings incorporate a semi-private central courtyard. Finally, the analysis reveals that there are no transitional spaces between private property and public space (see Figure 2.10).

b *Building height and façade composition.* The vast majority of the edifices around the boulevard (93.85 per cent) are two or three floors high, while 4.60 per cent have four floors. The remaining percentage is single-storey buildings. In consequence, the building height around the boulevard allows natural control from the interior to the exterior (Figure 2.10).

c *Land use and building occupation.* The analysis shows that there is a direct connection between spatial hierarchy and land use of the buildings: 56.92 per cent of the buildings have retail premises on the ground floor, while on the upper floor there is residential use. Thus most of the edifices around the boulevard present a residential mixed use. Of the buildings studied, 29.23 per cent are residential, 10.77 per cent are facilities, and 3 per cent are for retail use only.

d *Private and public area maintenance.* Photographs of the study area demonstrate that the maintenance of building façades and 24 de Mayo Boulevard is not optimal and they show signs of urban deterioration. Low aesthetic quality, according to CPTED theory, could increase the likelihood of

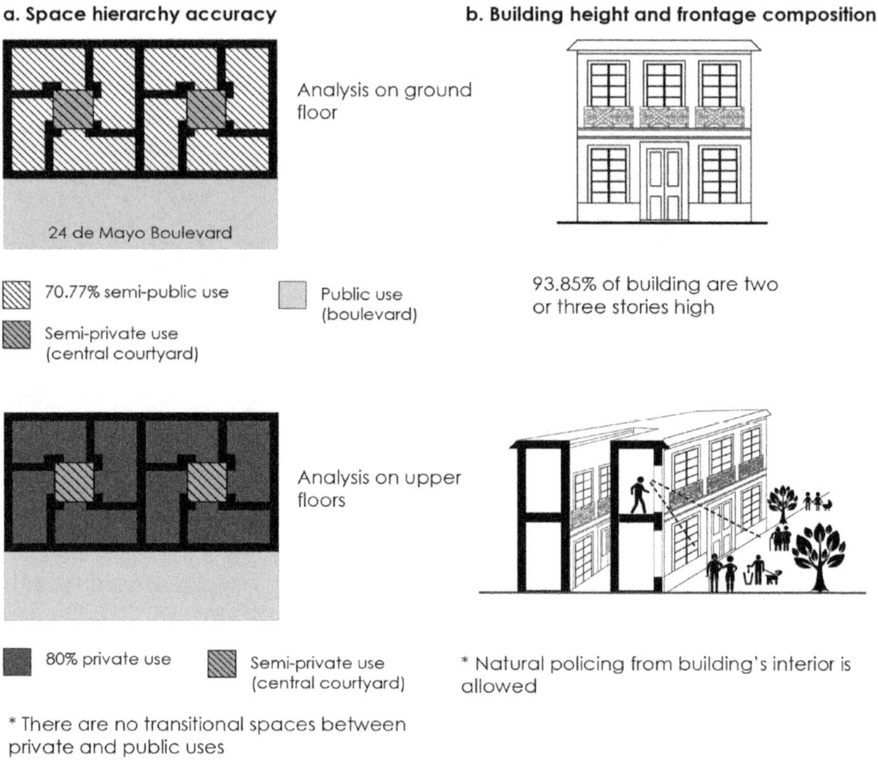

Figure 2.10 Architectural diagrams exploring space hierarchy accuracy, building height, and frontage composition on 24 de Mayo Boulevard.

criminal acts, as less concern and control indicate a greater tolerance of disorder (Crowe and Fennelly 2013: 28) (see Figure 2.11).

- *Number of access points.* More than 70 per cent of the buildings have direct access to the boulevard. A small percentage (7.70 per cent) has two access points and almost 20 per cent of the buildings have three or more.

Guayaquil's historic centre

The architecture of the historic centre in Guayaquil is characterized by an arcade typological model. The arcade is a covered area located on the ground floor of a building; it is privately owned but generally used for public circulation. It is developed inside the urban boundary of the building, with the construction of pillars acting as a physical limit of the public space (Santana Moncayo 2015: 681). In addition to public mobility, the arcades provide comfortable shelter in high temperatures, humid weather, and rain, which allows free pedestrian circulation throughout the year.

Figure 2.11 Low aesthetic quality in private and public areas on the boulevard in Quito.

In Ecuador the origin of this archetype comes from rural areas where residential units were designed to avoid floods and provide protection for farm animals. In coastal zones these covered spaces were also used to keep boats and fishing gear. This spatial configuration was transferred to the cities, and nowadays the arcade offers inhabitants space for meetings and retail activities (ibid.: 682).

The spatial structure of this architectural model implies a continuous pedestrian circulation on the ground level that connects the lower floors of the buildings and provides space for retail activities during the day. It is important to recognize that the functional structure of this architectural model has been transformed over time. It is common to use the arcades as expansion for the retail spaces, creating active street fronts that are beneficial for the natural surveillance of the neighbourhood.

On the other hand, the spatial characteristics combined with ambiguity of use (a private space for public use) generate a lack of clarity in terms of space hierarchy accuracy. According to Santana Moncayo (ibid.: 687), the consequences of this ambiguity of use is that the portals are often considered as hiding places for citizens with criminal intentions or are used as public toilets. At the neighbourhood scale, the arcade is also considered as a space organizer that extends the activities from the interior to the public space and, at the same time, ensures a continuous pedestrian flow in spite of the tough weather conditions of the coastal city (see Figure 2.12).

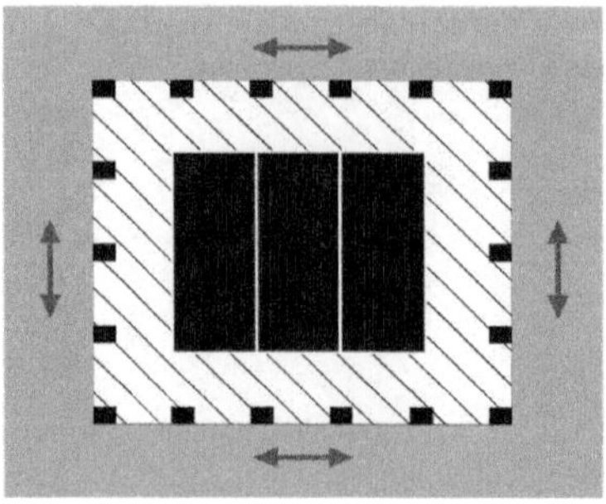

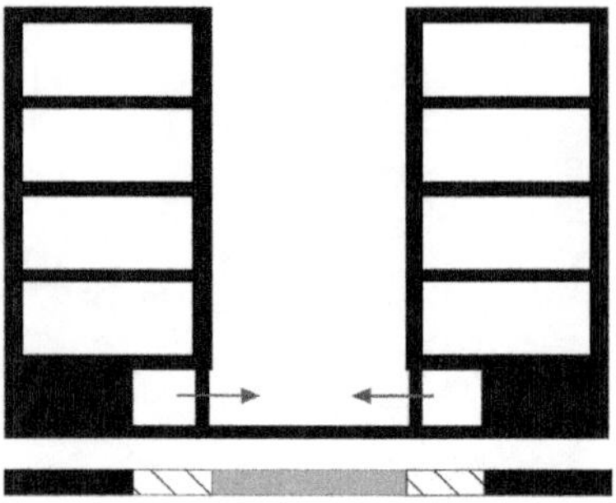

Figure 2.12 Typological abstraction – architectural diagram of Guayaquil.

Finally, local policies recognize the arcades as private places with public use, and emphasize that the maintenance responsibility lies with the owners of the premises of the ground floors (ibid.: 687). In addition, two public policies could influence the street vitality of the zone. First, the owners of retail premises on the ground floor are authorized to place mobile street furniture (such as chairs and tables) in the arcade if the municipal authorities have issued a yearly permit for such activities (Muy Ilustre Municipio de Guayaquil 2011: 6). Second, there is a ban on informal sales with mobile kiosks in the historic centre (Muy Ilustre Municipio de Guayaquil 2006: 3).

As with 24 de Mayo Boulevard in Quito, an analysis of the spatial strategies for psychological safety on Simón Bolívar Waterfront is as follows.

a *Linearity.* A single continuous façade marks the linearity of the waterfront, since the public space is limited to the east side by the Guayas River. In addition, between the arcades of the buildings and the river promenade the presence of the four-lane Malecón Avenue constitutes a physical barrier that decreases the linearity and continuity of the area. Finally, the linearity of the space is completely interrupted by the presence of a two-metre-high fence surrounding the public space and transforming it into a semi-public space (see Figure 2.13).

b *Number of secondary access points.* One-third of the buildings have secondary access points, yet the entries do not face small alleys but two-lane streets. Moreover, the public space itself has six entrances facing the avenue, distributed approximately one every block. Finally, 67 per cent of the buildings possess single entrances, which contributes to the general sense of safety in the area (see Figure 2.13).

c *Synergy.* In terms of synergy, the presence of the Malecón Simón Bolívar Avenue isolates the public space and prevents users from perceiving the

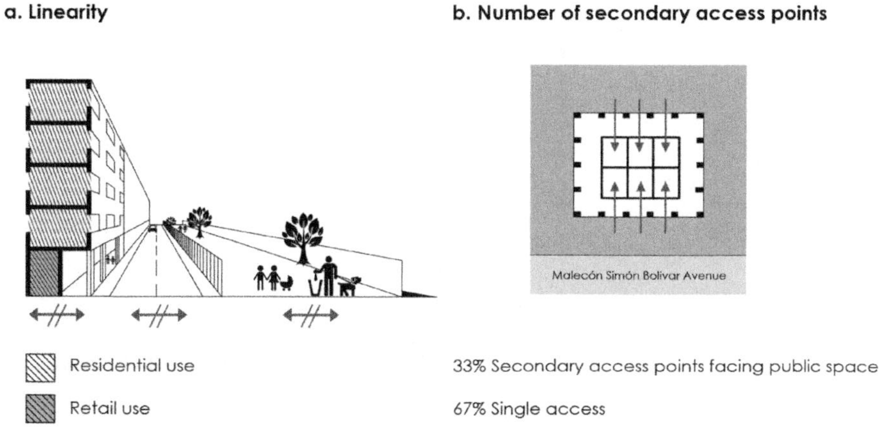

Figure 2.13 Analysis of linearity and number of secondary access points on Simón Bolívar Waterfront in Guayaquil.

different urban elements as parts of a global spatial configuration. The public space is designed as a heterogeneous district separated from the one formed by the blocks located in the west area. In the same way, there is a rupture between the waterfront strip and the river due to the separation by railings and platforms above the river. Finally, the psychological safety factor is negatively affected by the division of the area into three areas that are perceived as heterogeneous districts, preventing users from perceiving the area as a single urban structure (see Figure 2.14).

d *Connectivity.* In terms of the connectivity of the street network, 27 street segments were analysed and the results show that 26 per cent have the highest level of connectivity, level 6. In addition, 40.7 per cent present a connectivity level of 5, and less than 34 per cent of the segments are in level 4, which is still considered an adequate level in terms of connectivity within a place. There are no dead ends and the blocks are not longer than 100 metres, which contributes to the walkability and legibility of the area (see Figure 2.15).

e *Visual connection.* The psychological sense of safety in the area is crucially influenced by the presence of visual barriers that limit natural control of the area. As stated before, the area has three districts separated from each other by physical or perceptual barriers. In spite of the public character of the ground floors, the inhabitants have opted to close the visual connection with the main avenue: 22.72 per cent of the buildings analysed have physical barriers that prevent the natural connection with the Malecón Avenue. In the same way, inside the public space a user would find a myriad of visual impediments reducing the relation between districts (see Figure 2.16).

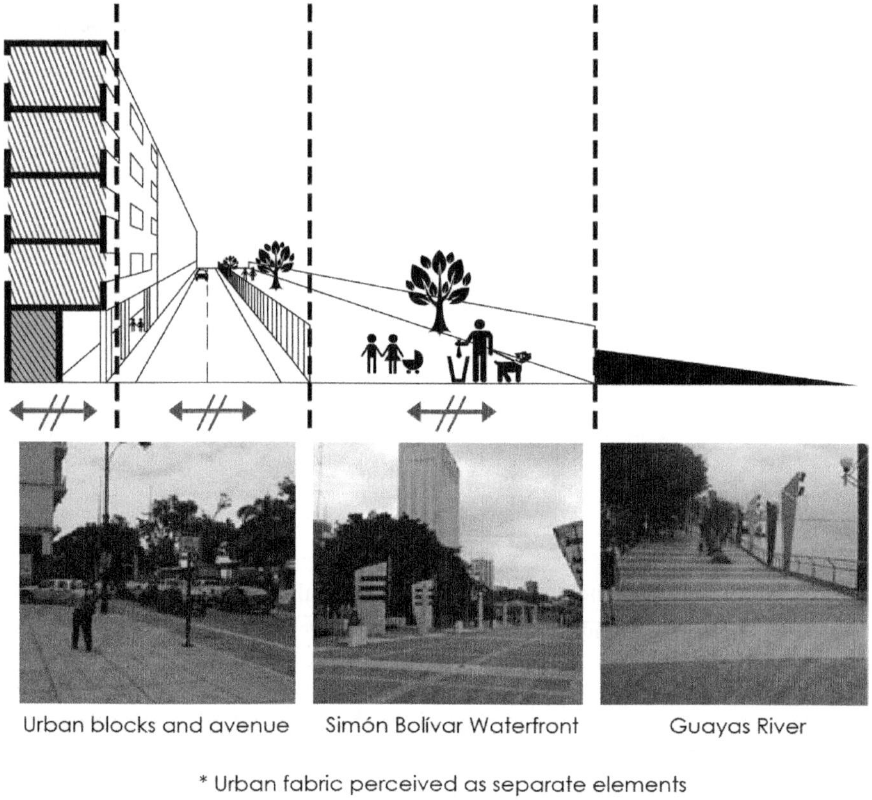

Figure 2.14 Perception of urban elements on Simón Bolívar Waterfront.

In terms of defence for safety, and again following the spatial strategies described earlier, the analysis on the Simón Bolívar Waterfront is as follows.

a *The accuracy of spatial hierarchy.* More than 50 per cent of the analysed buildings have semi-public use on the ground floor, due to the gallery that allows pedestrian circulation. The transitional space formed in the arcades allows retail establishments to develop, which enables semi-public use. On the ground floor, 13.64 per cent of the buildings have semi-private use and the remaining 31.82 per cent are private use. It is important to highlight that three lots in the area are used as parking spaces, which also modifies the hierarchy of the space due to the semi-private use. Conversely, for the upper floors only 22.72 per cent of the buildings are semi-publicly used, while 63.64 per cent are completely private. The remaining 13.65 per cent are considered as semi-private (see Figure 2.17).

d. Connectivity

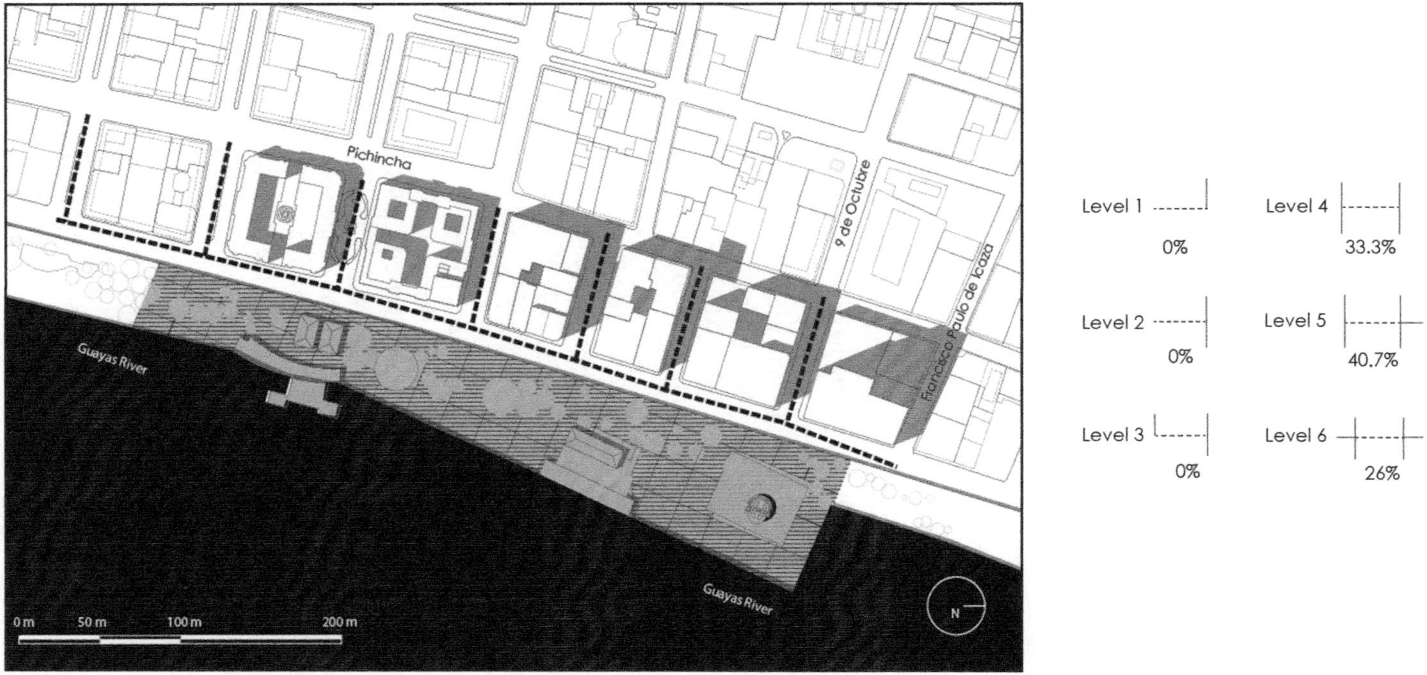

Figure 2.15 Connectivity study of Simón Bolívar Waterfront.

Figure 2.16 Visual impediments on the waterfront.

b *Building height and façade composition.* The skyline of the waterfront reveals that 90.91 per cent of buildings are high-rise with nine floors on average, while only 9.09 per cent have fewer than three stories. This spatial characteristic prevents direct contact of the inhabitants of the upper floors (fourth and higher) with the activities of the waterfront, which reduces considerably the natural surveillance of the place according to theories of defensible space (see Figure 2.17). Moreover, this phenomenon, combined with visual impediments, contributes to the separation of the waterfront from the city.

c *Land use and building occupation.* Mixed use is found in 27.27 per cent of the buildings, with retail activities on the ground floor and residences on the upper floors. In addition, 36.36 per cent function as public facilities, 22.73 per cent are dedicated to office use, and 13.64 per cent are unused, abandoned buildings.

d *Private and public area maintenance.* The study shows that only 31.82 per cent of the buildings maintain their façades and the circulation corridor in the arcades. The majority of the façades that are properly maintained are on buildings privately owned by banks or the government sector which rely on a good corporate image to attract customers to the interior. The rest of the buildings show a lack of maintenance in the façades facing the public space, preventing a unified imagery of the neighbourhood (see Figure 2.18).

e *Number of access points.* There are single entrances in 68.10 per cent of the buildings, almost 20 per cent have two entries, and 13.64 per cent have more than three access points to the upper floors. This spatial characteristic could contribute to reinforcing the vitality of the public space outside the buildings; however, most entrances are guarded by private security companies, preventing active pedestrian flow from the public space to the private core.

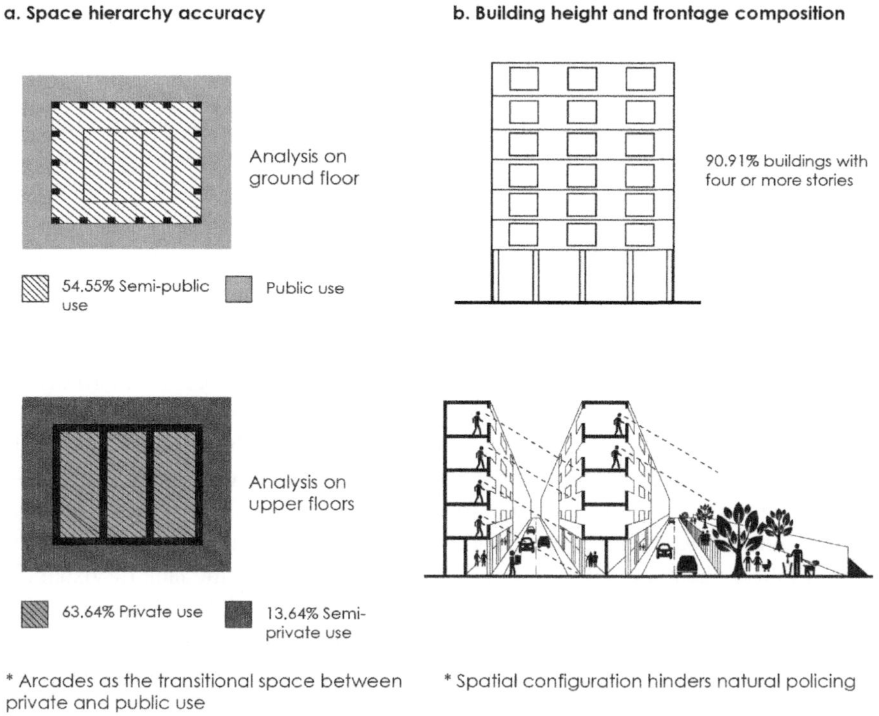

Figure 2.17 Architectural diagrams exploring space hierarchy accuracy, building height, and frontage composition around the waterfront in Guayaquil.

Figure 2.18 Low aesthetic quality in private areas and high maintenance in public areas in Guayaquil.

Conclusions

This concluding section aims to draw together the key observations from the case studies and make several reflections about how they link to broader discussions in urban studies. Overall, the chapter emphasizes the importance of spatial configuration at the neighbourhood level in fostering effective urban safety and crime prevention approaches. The chapter develops a conceptual framework for spatial strategies for urban safety and crime prevention that builds on models favouring either enclosure or encounter approaches to urban design, and the respective foundational work on these approaches by Jane Jacobs (1961) and Oscar Newman (1972).

In Quito the central courtyard typology defines a clear separation between private and public space. Interior spaces are protected, while public places are considered encounter areas between inhabitants and strangers. Furthermore, the approach fosters a sense of territoriality and natural policing through the combination of land use, building height, and access control. The spatial configuration in Quito's historic centre fosters the prevention of crime, while at the neighbourhood level the configuration of the urban fabric nurtures a sense of psychological safety due to the linear configuration of the space, the synergy existing in the urban fabric, the degree of connectivity of the streets, and the lack of visual obstacles in public places. The case study shows that the spatial configuration of the buildings around 24 de Mayo Boulevard defines the boundaries of space for private life and eases natural control of the public space in terms of urban safety and crime prevention.

In contrast to Quito, Guayaquil's spatial configuration presents an ambiguity of uses. Private and public spaces are not clearly defined due to the transitional circulation through the arcades. Prior to the configuration of the modern city, this approach fostered social and commercial relations. In contemporary settings, however, these characteristics are lost due to current local policies that require yearly permits and limit spontaneous activities in the arcades. In addition, the lack of accuracy in terms of space hierarchy, building height, and the presence of vacant lots hinders a sense of unified territoriality of the neighbourhood. An additional compounding factor is that inhabitants have modified the existing spatial configuration – by sealing windows or doors, for instance – to acquire control and privacy of the buildings' interior.

Furthermore, the Guayaquil case study shows that the urban fabric is divided into three heterogeneous spaces: the arcades with the avenue, the waterfront, and the river. These divisions prevent inhabitants and users from perceiving the area as a single space, and create individual sections that transform the public space into a semi-public area with restrictions in terms of access and activity. In this context, natural surveillance to foster crime prevention is limited, resulting in lower levels of psychological safety and overall lower levels of urban safety.

These two case studies connect to the broader literature in urban studies on psychological safety and defence for safety in terms of design strategies. This distinction is useful, because it puts forward the ways in which ambiguities in the safety and security terminology manifest themselves in the spatial configuration of a city and in the behaviour of users in public places and surrounding streets. Like the broken windows theory, where the tolerance of minor incivilities could lead to urban decay, contemporary urban thinkers argue that 'people come where people are' (Gehl 2011: 25). That is to say, places where more people would feel welcome are those in which inhabitants and users willingly interact – a characteristic that sets the difference between design for enclosure and design for engagement.

What is more, developments in urban studies have emphasized four safety factors in the design of urban space, described in this chapter as psychological safety, defence for safety, behavioural safety, and safety against disaster. The chapter argues that urban and architectural form only directly influences the first two factors. It is crucial to recognize the spatial strategies that define both approaches and the connection with spatial configuration and its formal structure to design 'for safety' in relation to a specific context. In other words, spatial design can favour conditions that make people feel safer or prevent crime, but it does not limit criminal behaviour as such.

Finally, the chapter notes that the culturally accepted meaning of *seguridad* in Spanish is arguably one factor that prevents architects and urban designers in Latin America from approaching urban safety in a holistic way. The concept of *seguridad* mainly focuses on strategies to protect inhabitants against crime, and is therefore less sensitive to urban safety and crime prevention approaches. A more comprehensive approach to urban design would consider all four factors that intervene in urban safety and build a stronger empirical evidence base through additional case studies. Focusing only on strategies that protect inhabitants against crime could privilege a deterministic approach to urban design in which crime is not considered a social product that emerges from social inequality, but a result that emerges directly from the built environment. Based on the observations from the case studies, this chapter highlights that it is possible to go beyond such determinism towards a comprehensive approach to spatial configuration, urban safety, and crime prevention. The encounter approach discussed in this chapter also illustrates that there is a tried and tested way of designing urban space that can help contribute to the building of peaceful and inclusive societies in urban spaces.

References

Cai, K. and J. Wang. 2009. 'Urban Design Based on Public Safety: Discussion on Safety-Based Urban Design', *Frontiers of Architecture and Civil Engineering in China* 3(2), 219–227.

Cara, L. and C. López Rueda. 2017. 'Seguridad Urbana y Prevención del Crimen: Estrategias Urbanas y Arquitectónicas en el Centro Histórico de Quito', *Revista CienciAmérica* 6(3), 96–101.

Carmona, M., S. Tiesdell, T. Heath, and T. Oc. 2010. *Public Places – Urban Spaces: The Dimension of Urban Design*, 2nd edn. Oxford: Architectural Press.

Chiriboga, H. 2006. *Regeneración Urbana: Privatización del Espacio Público, Políticas de Seguridad y Tematización en Diario el Universo de Guayaquil*, Revista Académica de la Federación Latinoamérica de facultades de comunicación social. Guayaquil: Universidad Católica de Santiago de Guayaquil, available at http://dialogosfelafacs.net/wp-content/uploads/2015/75/75-revista-dialogos-regeneracion-urbana-privatizacion-del-espacio-publico.pdf (accessed 7 June 2018).

Clarke, R. 1980. '"Situational" Crime Prevention: Theory and Practice', *British Journal of Criminology* 20(2), 136–147.

Clarke, R. 1997. *Situational Crime Prevention: Successful Case Studies*. New York: Harrow and Heston.

Cozens, P. 2008. 'Crime Prevention Through Environmental Design', in R. Wortley and L. Mazerolle (eds) *Environmental Criminology and Crime Analysis*. Cullompton: Willan, 153–177.

Crowe, T. and L. Fennelly. 2013. *Crime Prevention through Environmental Design*. Oxford: Butterworth-Heinemann.

Dovey, K. 2000. 'Redistributing Danger: Enclosure and Encounter in Urban Design', *Australian Planner* 37(1), 10–13.

Dovey, K. 2016. *Urban Design Thinking: A Conceptual Toolkit*. London: Bloomsbury Academic.

Escudero López, E. 2016. 'Espacio Público y Seguridad', *Planur-e* 10 (Winter), available at www.planur-e.es/articulos/ver/espacio-p-blico-y-seguridad/completo (accessed 6 July 2018).

Garcés, C. 2004. 'Exclusión Constitutiva: Las Organizaciones Pantalla y lo Anti-Social en la Renovación Urbana de Guayaquil', *Iconos – Revista de Ciencias Sociales* 20, 53–63.

Geason, S. and P. Wilson. 1989. *Design out Crime: Crime Prevention Through Environmental Design*, Crime Prevention Series. Canberra: Australian Institute of Criminology, available at https://aic.gov.au/publications/crimprev/cpted (accessed 6 July 2018).

Gehl, J. 2011. *Life Between Buildings*. London: Island Press.

Glaeser, E. 2012. *Triumph of the Cities*. London: Pan Books.

Hillier, B. 1996. *Space Is the Machine: A Configurational Theory of Architecture*. London: Press Syndicate of the University of Cambridge.

Hillier, B. and J. Hanson. 1984. *The Social Logic of Space*. Cambridge: Cambridge University Press.

Hillier, B. and J. Hanson. 2004. 'Can Streets Be Made Safe?', *Urban Design International* 9(1), 31–45.

Jacobs, J. 1961. *The Death and Life of Great American Cities*. New York: Random House.

Lang, J. 2005a. 'Imagery and Habitability in Urban Design: Can These Two Functions Be Reconciled?', *City Planning Review* 2, 27–32.

Lang, J. 2005b. *Urban Design: A Typology of Procedures and Products*. Oxford: Architectural Press.

Lazarus R. S. and J. B. Cohen. 1977. 'Environmental Stress', in I. Altman and J. F. Wohlwill (eds) *Human Behavior and Environment*. Boston, MA: Springer, 89–127.

Lefebvre, H. 2007 [1974]. *The Production of Space*. Oxford: Blackwell Publishers.

Lynch, K. 1960. *The Image of the City*. Cambridge, MA: MIT Press.

Martí Arís, C. 2014. *Las Variaciones de la Identidad: Ensayo Sobre el Tipo en Arquitectura*. Barcelona: Arquia, Fundación Caja de Arquitectos.

Maslow, A. 1987. *Motivation and Personality*. New York: Harper and Row.

Micara, J. D. 1985. *Architettura e Spazi dell'Islam: Le Isituzioni Collettive e la Vita Urbana*. Rome: Carucci.

Minton, A. 2012. *Ground Control*. London: Penguin Books.

Muy Ilustre Municipio de Guayaquil. 2006. 'Ordenanza Sustitutiva de la Ordenanza que Norma la Instalación de Kioskos y Carretillas.' Guayaquil: Gaceta Oficial.

Muy Ilustre Municipio de Guayaquil. 2011. 'Ordenanza que Norma la Utilización de Mesas y Sillas en Espacios Públicos.' Guayaquil: Gaceta Oficial.

Newman, O. 1972. *Defensible Space: Crime Prevention Through Urban Design*. New York: Macmillan.

Newman, O. 1996. *Creating Defensible Space*. Washington, DC: US Department of Housing and Urban Development, Office of Policy Development and Research.

Ortega, J. F. 2014. *Regeneración Urbana: Ddiscurso Patrimonial Oficial y Segregación Social en la Av. 24 de Mayo*. Quito: Universidad Andina Simón Bolívar.

Palimariciuc, L. 2016. *Urbanismo y Seguiridad. Verificación del Método CPTED en un Barrio de Madrid*. Madrid: Escuela Técnica Superior de Arquitectura de Madrid, available at http://oa.upm.es/39235/ (accessed 6 July 2018).

Raco, M. 2007. 'Securing Sustainable Communities: Citizenship, Safety and Sustainability in the New Urban Planning', *European Urban and Regional Studies* 14(4), 305–320.

Rossi, A. 1984. *The Architecture of the City*. Cambridge, MA: MIT Press.

Real Academia Española. 2014. 'Seguridad Ciudadana', in *Diccionario de la Lengua Española* (23rd edn). Madrid: Real Academia Española, available at http://dle.rae.es/?id=XTrIaQd (accessed 6 July 2018).

Santana Moncayo, C. A. 2015. 'Guayaquil, Ciudad de Soportales: Una Reflexión Acerca de Su Importancia y Su Uso Actual', *Pasos* 13(3), 681–696.

Silva, M. B. 2001. 'La Vivienda a Patios de Origen Hispánico y Su Difusión en Iberoamérica', in *Actas III Congreso Internacional del Barroco Americano: Territorio, Arte, Espacio y Sociedad*. Seville: Universidad Pablo de Olavide, 875–896.

Wong, C. 2005. 'Del Caos al Orden: Guayaquil y Su Desarrollo Urbano Actual', *Ciudades* 9, 179–191.

3 Gangland *terra nullius*

Violence, territoriality, and bottom-up spaces of peacebuilding in urban Nicaragua

Dennis Rodgers

Introduction

Gangs are ubiquitously considered primary sources of violence and insecurity across the world (Hazen and Rodgers 2014). This is currently perhaps most obviously the case in post-Cold War Central America, where gangs, known variably as *maras* and *pandillas*, are widely perceived as the most important actors within a contemporary panorama of rampant criminality characterized by levels of violence often surpassing those of the revolutionary conflicts that affected the region during the 1970s and 1980s (Rodgers 2009). Not surprisingly, perhaps, Central American gangs have been portrayed as a critical regional security threat, and the corresponding policy response has been brutally repressive, to the extent that it is no exaggeration to talk of Central American states having declared a veritable 'war on gangs' in recent years (Jütersonke *et al.* 2009; Wolf 2017; Fontes 2018).

Given that gangs are predominantly an urban phenomenon – most likely due to the fact that there needs to be a 'critical demographic mass' of youth for a gang to emerge within any given locality (Fischer 1975; Rodgers 2010) – this war has not surprisingly been waged principally in the isthmus' cities. Certainly, Central American gangs are often portrayed as a 'new urban insurgency' aiming 'to depose or control the governments of targeted countries' through 'coups d'street' (Manwaring 2005: 2), and much of public action against gangs has involved attempting to 'secure' urban territory and render it 'safe' (Gutiérrez Rivera 2010, 2013). The logic behind this is that 'territoriality is generally considered a defining characteristic of youth gangs' (Moore *et al.* 1983: 183), and much gang violence is perceived as revolving around their spatially controlling and dominating space.

This is widely considered to have critical consequences for the broader morphology of urban settlements, in so far as gang-affected cities often become 'splintered' (see Graham and Marvin 2001) into a patchwork quilt of disconnected 'no-go' areas, with rival gangs battling over control and domination and the broader mass of the urban population constituting 'collateral damage'. As Murray (2017) points out more generally, this kind of 'urbanism of exception' fundamentally undermines the agglomerative benefits of city living and perpetrates seemingly never-ending cycles of violence, thereby making gangs and their 'ganglands' – to use the term that Frederic Thrasher (1936 [1927]) famously coined in his pioneering study of gangs in 1920s Chicago – key vectors of urban conflict and insecurity.

At the same time, however, such apocalyptic visions are arguably based on a highly monolithic and indeed profoundly misguided understanding of the spatial and sovereign logic of both gangs and ganglands. Drawing on ethnographic research on gang dynamics carried out in *barrio* Luis Fanor Hernández, a poor neighbourhood in Managua, Nicaragua, this chapter aims first to demonstrate how gang violence can often be more constitutive than destructive, and second to offer an alternative conception of gangland that highlights its variegated nature as well as that of gang authority more generally (see Rodgers 2007a and Rodgers 2019 for methodological details). In doing so, it offers something of a counterintuitive response to the volume's wider concern with 'urban safety' as a way of particular thinking about the spatial configurations of cities – crucially by identifying potential peacebuilding avenues within what at first glance would seem to be a space whose dynamics are unmitigatedly and wholly violent.

Gangland political economy

According to Thrasher (1936 [1927]: 22), ganglands are 'geographically and socially' distinct types of urban area with the particular characteristic that their spatial organization fundamentally responds to gang dynamics. More specifically, Thrasher argued that a gangland was

> … medieval and feudal in its organization rather than urban and modern [like the rest of the city]. The hang-out of the gang is its castle and the center of a feudal estate which it guards jealously. Gang leaders hold sway like barons of old, watchful of invaders and ready to swoop down upon the lands of rivals and carry off booty or prisoners or to inflict punishment upon their enemies.
>
> (Ibid.: 6)

As a result, Thrasher argued that a gangland was marked by 'continuous … disorder and violence …so pronounced as to give the impression that [it was] … like a frontier … a "no man's land", lawless, godless, wild' (ibid.: 5–6), which he contended reinforced the marginality of ganglands within the broader Chicagoan urban panorama.

Numerous studies have subsequently echoed Thrasher, whether in relation to gangs in Chicago (Keiser 1969; Suttles 1968), other US cities (Adamson 1998; Moore *et al.* 1983; Tita *et al.* 2005; Whyte 1993 [1943]), or other parts of the world, including Russia (Salagaev and Safin 2014; Stephenson 2015), France (Lepoutre 1997; Mohammed 2011), the UK (Kintrea *et al.* 2010; Pickering *et al.* 2012), India (Sen 2014), South Africa (Glaser 1998; Jensen 2008), the Democratic Republic of the Congo (Geenen 2009), Nicaragua (Rodgers 2006, 2009;

Rocha 2007), and Honduras (Gutiérrez Rivera 2010, 2013), to mention just a few examples. While frequently offering extremely useful insights into gang dynamics, most such studies tend to assume a certain singularity to a gang's territoriality. Even if there is some recognition that gang territories are highly contingent and fluctuate both with conflict and as gangs emerge and disappear over time (Jensen 2008: 81–82), and also that gang territory 'borderlands' can potentially constitute ambiguous 'grey areas' (Moore *et al.* 1983: 184–185), most studies tend to assume that a gang's territory will be a relatively clearly bounded space, and that the gang will hold sway over it in a uniform manner.

In this latter regard, a commonly repeated element of the above studies is that gangs promote a form of order using particular practices of territorial regulation and control – whether in the absence of or in collaboration with other forms of authority such as the state (see Arias 2006) – based on violence, including most notably intergang conflict and differential patterns of victimization. Certainly, in my own work on gangs in Nicaragua, for example, I have explicitly explored how gangs promoted a specific form of local socio-spatial ordering. I label this a form of 'social sovereignty' (Rodgers 2006: 289), in contrast to state-based sovereign authority, and have particularly explored how the gang's sovereign regime can provide non-gang member residents within the territory over which it holds sway with certain benefits (as well as drawbacks). In particular, I show how much of the gang's violence, which at first glance might seem anarchic and highly deleterious, in fact follows a definite logic aimed at creating order rather than chaos, or, in other words, how it is socially constitutive rather than solely destructive (see Rodgers ibid., 2007c, 2015, 2017).

This was perhaps most counterintuitively the case with respect to the gang warfare in which these gangs engaged. Although gang wars often had extremely negative consequences for local neighbourhood populations, these were generally indirect, with the threat stemming principally from other gangs with whom the local gang engaged in a prescribed manner that offered local communities a form of order and predictability. Certainly, gang warfare rigidly obeyed a number of precise rules and practices, to the extent that it can be qualified as having been 'semi-ritualized'. In particular, the first battle of a gang war typically involved fighting with fists and stones, but each new battle involved an escalation of weaponry, first to sticks, then to knives and broken bottles, and eventually to mortars, guns, and AK-47s. Although the rate of escalation varied, its sequence never did – i.e. gangs did not begin their wars immediately with firearms. The fixed nature of gang warfare constituted something of a restraining mechanism, in so far as escalation is a process in which each stage calls for a greater but definite intensity of action and is therefore always under actors' control. But it also provided local neighbourhood inhabitants with an 'early warning system', thereby offering a means of circumscribing the 'all-pervading unpredictability' of violence (Arendt 1969: 5).

This positive function was widely recognized and indeed appreciated by local inhabitants, who not only frequently talked very approvingly about their local gang but also often provided assistance to its members, for example hiding them if they were chased into the neighbourhood by other gang members or private security guards while engaging in delinquent activities outside the neighbourhood. Gang members returned the favour by never targeting members of their local community when they engaged in delinquency, actively going out of their way to protect local neighbourhood inhabitants whenever they saw them being threatened by outsiders, and frequently providing (free) bodyguard services as well as watching out for people's property. The motivation offered by gang members in *barrio* Luis Fanor Hernández for such practices was that they wanted to show their 'love' – literally, *querer* – for their local neighbourhood. 'We show our love for the neighbourhood by fighting other gangs', a gang member called Miguel claimed, for example, while another called Julio explained in more detail that 'you show the neighbourhood that you love it by putting yourself in danger for people, by protecting them from other gangs.... You look after the neighbourhood in that way, you help them, keep them safe' (see Rodgers 2006 for more details).

As such, the gang can be said to have constituted something of a functional source of security for local inhabitants, which in the post-war context of political polarization, disillusion, and chronic insecurity that characterized Nicaragua at the time (see Rodgers 2008, 2014) was something that was explicitly recognized and even celebrated. Certainly, it was common to hear the expression *la pandilla es el barrio* (the gang is the neighbourhood) used by both gang members and non-gang members in *barrio* Luis Fanor Hernández. Indeed, this went even further than the gang providing physical security, insofar as there was also a clear and widespread sense of identification with the local gang and its highly performative violence; this was evident in the 'aesthetic pleasure' (Bloch 1996: 216) that local inhabitants derived from swapping stories about the gang, including in particular exchanging eye-witness accounts of dramatic acts by gang members during warfare, and spreading rumours and retelling various incidents over and over again. As such, the gang can be said to have constituted something of an institutional medium for the symbolic constitution of a sense of communal collective identity in *barrio* Luis Fanor Hernández, which was otherwise lacking in the neighbourhood (see Rodgers 2007b).

The variegated spaces of gangland

While I stand by the above analysis of the gang's ordering functions, I have perhaps been guilty of failing to consider properly the internal variation in gangland territorial dynamics. In particular, it is clear that diverse types of gang-related sovereignty regimes can coexist within one gangland territory, and that variety exists within Nicaraguan ganglands. This is something that became retrospectively clear to me when I recently reconsidered my notes concerning an event that occurred in April 1997, while I was running an errand with a gang member called Julio at the Huembes market near *barrio* Luis Fanor Hernández. As we weaved our way between the market stalls, making our way to the food hall to do some shopping for an elderly inhabitant of *barrio* Luis Fanor

Hernández who was too poorly to shop for herself, I suddenly spotted four gang members from another neighbourhood, *barrio* Nosara, with whom the *barrio* Luis Fanor Hernández gang was in conflict….

> 'Shit, *maje*', I hissed to Julio, 'let's get the fuck out of here, there's four *pandilleros* from *barrio* Nosara coming our way and they've all got machetes and we're unarmed!'
>
> '*Calmaté*, Dennis', he answered, 'it's fine, we're in the market, they're not going to do anything to us.'
>
> 'What do you mean? We're at fucking war with them!'
>
> 'Sure, but wars are only fought on gang territory, here in the market it's different, it's nobody's land, you only attack somebody if you've got a *traido* (personal vendetta) with them, and I don't have any beef with any of those four guys, and nor do you.'
>
> Julio then proceeded to confidently walk towards the *barrio* Nosara gang members, while I trailed behind him somewhat diffidently, and sure enough, as we passed by them, one gave us a curt nod while the others ignored us, and we peacefully went our separate ways.
>
> 'I don't get it, Julio', I said to him afterwards, 'I thought that when we were at war with another gang, we were always at war with them. I know that this can change from one war to the next, and yesterday's enemy can be tomorrow's ally, but we normally never let rival gang members go by unchallenged when we're at war with them, so what was that all about in the market?'
>
> '*Pues*, Dennis', he answered patiently, 'it depends where you meet them. If they came into the *barrio*, then of course we'd beat the shit out of them and throw them out, it's our heartland, and if we'd met them in *colonia* Las Condes that would have been different too, as that's part of our operating territory, and you can't allow rival gang members to be there unchallenged if you come across them there. But the market is nobody's territory, so it's not the same. Unless you have a personal gripe against somebody, you just live and let live here, anybody can go through or be there at the same time, war or no war….'

This episode clearly suggests that, contrary to what is often assumed, gangland space can be internally variegated. More specifically, Julio was highlighting at least three distinct types of territories within which the *barrio* Luis Fanor Hernández gang was present, namely the *barrio* (or what might be termed the gang's 'heartland'), its 'operating' territory, and the Huembes market. At first glance the difference between the three might seem to be principally spatial, especially as each had a clear geography. The 'heartland' territory was made up of the *barrio* Luis Fanor Hernández neighbourhood, a 300-metre stretch of a boundary *cauce* (floodwater channel), including its banks on both sides, a small wasteland that abutted the neighbourhood but ran into a neighbouring *asentamiento* (slum), and the whole of a pedestrian bridge over the *cauce*. The 'operating' territory of the gang included several nearby wastelands, major roadways, and thoroughfares, as well as *colonia* Las Condes, which did not have a rival gang, being a more affluent middle-class neighbourhood, and also the neighbourhoods of other gangs whom they had defeated in war. Finally, the Huembes was a formally constituted municipal space that covered a well-delimited location.

A closer consideration, however, reveals the distinction between these different territories to be less spatial than based on differences in the sovereignty regime exercised by the gang over them. The gang's sovereign claim over its heartland was unambiguous and universal, and involved more than just territorial claims but also a particular relationship with those inhabiting this space. Certainly, a significant amount of gang members' activities and violence related to the protection of local inhabitants, whom they never preyed upon. They explicitly behaved in this way towards the latter because of their residence within a definite bounded territorial unit with which the gang identified, namely the neighbourhood, and the gang enacted its sovereign claim through a mixture of active and passive territorially based strategies.

Gang members would, for example, spend significant chunks of the day and night sat in small groups on neighbourhood street corners, from where they could observe comings and goings in the neighbourhood; individuals would move from group to group regularly to compare notes, while others would explicitly engage in perimeter 'patrols' of the neighbourhood. Anybody who was not from the neighbourhood would be challenged, and unless they could justify their presence – by being able to link themselves to a local neighbourhood inhabitant – they were almost invariably beaten and chased out. As I describe elsewhere (Rodgers 2006), such behaviour was part of a broader set of strategies by which the gang enabled local inhabitants to live their lives a bit more securely – albeit within a restricted territorial perimeter – in a broader context characterized by high levels of chronic urban violence. The basis for the sovereign claim was unmitigatedly territorial, very much along the lines described by Gutíerrez Rivera (2010: 496): 'traditionally, gang territoriality has been understood as: the concentration of the gangs' activities within a "turf"; that this turf is relatively bounded; [and] that gang members defend the turf against invaders, usually the rival gang'.

The operating territory of the gang included areas where members carried out most of their delinquency, but these were clearly not considered part of the gang's territory in the same way as the heartland. This was especially obvious from the fact that the gang's sovereignty claim over its operating territory was both uneven and based on a different logic and strategy to that which it exercised in its heartland. In particular, *colonia* Las Condes was subject to a much more contingent sovereignty claim than *barrio* Luis Fanor Hernández.[1] This was partly because the *barrio* Luis Fanor Hernández gang did not have the numbers to regulate and control it in the same way as the heartland territory, and therefore other gangs circulated within it. Although the neighbourhood's proximity to *barrio* Luis Fanor Hernández – one of the entrances to the *barrio* came out of it

– meant that the gang could not 'just let anything happen there', as a gang member called Jader put it, and *barrio* Luis Fanor Hernández gang members would challenge rival gang members whenever they came across them in *colonia* Las Condes, they would not actively seek them out, nor did they patrol the *colonia*, which meant that in practice this neighbourhood was part of the operating territory of several different gangs.

This was not seen as a problem by gang members because the logic of the gang's sovereign claim over *colonia* Las Condes was not the same as that which it exercised over its heartland territory. Borrowing from Keiser's (1969: 27) insight in his study of the Vice Lords in Chicago in the 1960s that 'the distinction between section territories is based on differential responsibility rather than differential rights', it can be argued that the *barrio* Luis Fanor Hernández gang's sovereignty claim to its heartland was based on responsibility (to protect the neighbourhood), while its sovereignty claim to its operating territory was based on an assumed right (to prey upon its population). *Colonia* Las Condes was a resource for gang members, and the gang's violence was aimed principally at extraction, albeit discontinuous, as members would rob or mug people there whenever they needed to rather than on a regular basis. The gang did not need to control the *colonia*'s territory for this, but rather had to be able to prey on its population with impunity, including in particular being sure the residents would not call the authorities – something they achieved by being unpredictably and extremely violent in their predation, precipitating a climate of terror and fear among *colonia* Las Condes's inhabitants.

As Jader explained during a discussion in May 1997:

> … *pues*, you know, Dennis, Las Condes is where we go and rob a lot of the time, and so we've got to make sure that people there fear us so that they don't call the Police, because after a while, they get to know us, know who we are. That's why whenever you mug somebody there, you've got to hit them a couple of times more than you would normally, and tell them '*ojo*, I know where you live, so you'd better be careful', or else '*oye hijueputa*, you know who I am, I've got my eye on you always'. Sometimes we even go to Las Condes as a group, just to beat a few people up, and for the *Purísima*, you know, in December, when we go from house to house singing and asking treats, we always take a mortar or two and shoot into a couple of houses at random, so that people are scared. That way nobody ever calls the Police.

Seen from this perspective, instead of a territorial logic there was a biopolitical underpinning to the gang's sovereignty regime in *colonia* Las Condes – i.e. it focused first and foremost on disciplining a population rather than on controlling a space, contrary to the situation in the gang's heartland.

Finally, the *barrio* Luis Fanor Hernández gang made no sovereign claim to the Huembes market, which it considered 'nobody's territory', as Julio explained. In legal terms, the Huembes could arguably be conceived as a form of *terra nullius*. This notion, which derives from Roman law, is used in international law to describe territory that is not subject to the sovereignty of any state, or over which any prior sovereign has expressly or implicitly relinquished sovereignty. As Andrew Fitzmaurice (2008: 2–3) explained, the notion of '*terra nullius* is a product of the history of dispossession and the larger history of European expansion', and was used to justify the British conquest of Australia (see also Banner 2005) as well as the so-called 'scramble for Africa', although it arguably crystallized around late-nineteenth-century debates concerning the legal status of the polar regions, including especially 'islands that lay within the Arctic Circle and over which no sovereignty had been established'. Without wishing to get caught up in a tortuous debate about the way the concept of *terra nullius* has been used in practice, I suggest that it is heuristically applicable to the Huembes as a form of gangland space to distinguish it from other commonly-used labels such as 'no-man's land' or 'impersonal domains', which Gerard Suttles (1968) proposed in his famous study of gang territoriality in 1960s Chicago to describe two types of gangland areas that were not 'gang territory'.

The first type of gangland area constituted

> … areas being razed, places of industrial concentration, or expressways. All of them are sections where people do not live permanently and over which no one exercises a personal surveillance. Given local ideas about who an area can 'belong to', this creates a kind of social vacuum where the usual guarantees of social order and control are lacking. Ordinarily they are viewed as dangerous, and people cross them 'at their own risk'.
>
> (Ibid.: 35)

The second type consisted of

> … non-residential areas where the safety of a passer-by is mostly in the hands of impersonal authorities who are either acting on behalf of someone else's interests or whose responsibilities do not include the entire day. For the most part, these are major thoroughfares flanked by impersonal business establishments or public institutions. During the daytime these are patrolled fairly well by businessmen, bureaucrats, policemen, and customers. At night, however, … there is no street life, there are no lights in the windows, and there is almost no reason for anyone to be there.
>
> (Ibid.: 36)

In other words, the distinguishing criteria Suttles proposed for these spaces were respectively either abandonment or control by a more distant form of authority (the state), neither of which applies to the Huembes market. It was not an abandoned space – quite the contrary, as it was a thriving local market and gang members from multiple neighbourhoods often went there – and while the Nicaraguan state, or perhaps more accurately the Managua municipality, would theoretically have claimed to exercise sovereignty over the Huembes, it was very rarely present.

Indeed, in terms of local authority it was in fact gang members who often acted as the arbitrators of disputes between merchants, for example. While this at first glance might seem to represent an exercise of sovereignty by gangs, they did so on an ad hoc basis, and no gang claimed the Huembes as its territory or its population as a resource, making it very different from other gangland areas, hence my suggestion that it be conceived as *terra nullius*. I also suggest this explicitly in opposition to other potentially germane notions such as 'sanctuary space', as developed by Jon Wolseth (2011: 105) in relation to Pentecostal conversion, because these 'lift [gang members] out of the violence of the everyday and re-inscribe… them into a new and protective social space'. The Huembes did nothing of the sort, as the market was a space of delinquency and other forms of more individualized violence, including in particular working out individual *traidos* or personal vendettas (which frequently became spectator sports for bored vendors on slow business days). Similarly, the labelling of vast swathes of the African continent as *terra nullius* in the late nineteenth century was the basis upon which all-out war between imperial powers was arguably avoided, in so far as it established the idea that the continent was 'nobody's territory', and therefore had to be effectively occupied before being formally claimed. This did not mean, however, that skirmishes did not occur on a regular basis, for example as countries sent rival exploratory expeditions or fought over resource extraction.

Gangland peacebuilding

Conceived as *terra nullius*, the Huembes as described above arguably emerges as a space of exception within the broader gangland context, where different violent actors can coexist and engage with each other in ways that do not have to entail antagonism. This is important, because it clearly opens possibilities for concrete peacebuilding initiatives within gangland areas. Indeed, it was often in *terra nullius* spaces such as the Huembes that truces and the ends of gang wars were negotiated between the *barrio* Luis Fanor Hernández gang and its rivals, or more accurately between individual gang members acting as peacemaking agents for their respective gangs. These were often individuals with kinship links across the two gangs, something which was not necessarily uncommon insofar as gangs tended to fight against other local gangs. In one particular case, for example, two cousins who were members of different gangs were able to meet up peacefully in the context of a family celebration being held at a Huembes restaurant and discuss 'terms', so to speak. At one level such individuals were arguably analogous in structural terms to the 'leopard-skin chiefs' discussed by Evans-Pritchard (1940: 163–164) in relation to the Nuer of southern Sudan, who would often mediate between both warring parties and feuding individuals. While the latter's semi-sacred status allowed them to move between the territories or homesteads of rival parties or individuals, this was not the case of the cousins, who could only meet to talk at the Huembes due to its *terra nullius* status.

This is important, because as Cruz and Durán-Martínez (2016) point out, most gang-related peacebuilding initiatives, at least in Latin America, tend to fail, often because they lack an awareness of the spatial dimension of urban safety. Certainly, as Jones and Rodgers (2015) point out, the failure of Rio de Janeiro's (in)famous Urban Pacification programme, which involved the deployment of specialized Police Pacification Units (*Unidade de Polícia Pacíficadora*) in the city's poor neighbourhoods (*favelas*) with the explicit aim of wrestling territorial control from locally dominant drug gangs, largely rests on the fact that such interventions displaced rather than eliminated violence, partly because they involved only a limited number of neighbourhoods (see Hoelscher and Norheim-Martinsen 2014). In a related manner, Saunders-Hastings (2015) describes how the occupation of a gang-controlled neighbourhood in Guatemala City by the National Army did not lead to an overall reduction of violence but rather its 'invisiblization', partly due to the fact that the local gang's activity mainly involved extortion rather than drug dealing, and relied on the members being able to intimidate and control the local population rather than a specific territory.

Most starkly, one reason why the 'gang truce' in El Salvador recently collapsed is arguably because it was underpinned by a tendency to assume uniform gang territoriality and dynamics, failing to take advantage of the possibilities offered by the variegated nature of ganglands. Certainly the Salvadoran state ultimately insisted that gangs cease all their activities everywhere – rather than, for example, promoting a step-by-step approach whereby gangs might first withdraw from their operating territories into their heartlands, before tackling the very different heartland gang dynamics (see van der Borgh and Savenije 2016). Meaningful urban peacebuilding might well have been achieved in cities such as San Salvador had different types of territorially related gang violence been disaggregated and received separate interventions. Seen from this perspective, understanding and recognizing the varying natures of gangland are potentially key to reducing gang violence in cities around the world. It also invites us to think about the issue of urban safety more broadly, and how to promote peacebuilding initiatives in contexts of metropolitan insecurity and conflict.

Conclusion

This chapter aims to deconstruct the logic and make-up of Nicaraguan ganglands to identify how these can contain informal spaces of potential peacebuilding. In this sense it follows on directly from Chapter 1 in this volume by Austin and Jütersonke, and its semiotic focus on spatial 'meaning-making' in urban contexts, albeit offering a bottom-up instance of the interface between urban safety and peacebuilding they theorize. It could of course be argued that ganglands inevitably remain local in scope, and therefore their dynamics cannot be related to the broader urban experience. Certainly there is no doubt that a gang's territorial reach is often quite limited. As Whyte pointed out in his classic study, *Street Corner Society*, 'the life of the corner boy proceeds along regular and narrowly circumscribed channels' (Whyte 1993 [1943]: 256), which means a well-defined and relatively limited space. As 'Doc', Whyte's central informant, pointed out:

> Fellows around here don't know what to do except within a radius of about three hundred yards. That's the truth, Bill. They come home from work, hang on the corner, go up to eat, back on the corner, up a show, and they come back to hang on the corner. If they're not on the corner, it's likely the boys will know where you can find them. Most of them stick to one corner. It's only rarely that a fellow will change his corner.
>
> (Ibid.)

Similarly, Tita *et al.* (2005: 293) noted how the Pittsburgh gangs they studied 'hang out in small, well-defined areas within neighborhoods', which they denominate as 'set space'.

Seen from this perspective, the existence of gangland *terra nullius* might seem to have limited peacebuilding potential. Such a viewpoint, however, ignores the fact that gangs are simultaneously phenomena and epiphenomena, or in other words, although they are autonomous local social institutions with complex internal logics and dynamics, they also fundamentally always reflect – and are shaped by – broader structural relations. Indeed, as Dimitriadis (2006: 338) points out, what made Thrasher's original study so innovative was precisely that it managed to capture both dimensions of gangs: it 'was highly contextual and relational, foregrounding the agency of young people in constructing their selves and social relations', but always doing so 'in their situational contexts'. This means that while Thrasher was able to provide extensive and detailed information about the minutiae of gangs, his study was not 'just' a gang study but also shed light on the political economy of Chicago. In particular, he explicitly situated ganglands within a broader understanding of the Chicagoan urban context, conceiving their particular dynamics as epitomizing the marginalization of particular socioeconomic groups – namely immigrants and African-Americans – and pointing to a much broader form of differentiated urban citizenship within American society at the time. Seen from this perspective, the variegated territorial regimes of ganglands and their particular dynamics might well indicate a way forward for tackling the deeper questions underpinning issues of urban insecurity in many cities across the world, offering a space through which to bring about concrete and meaningful peacebuilding from the bottom up.

Note

1 The same applied even more obviously to the *barrio* Luis Fanor Hernández gang's sovereignty claims over the territories of gangs which they defeated in war, as these were generally temporary, often lasting only a few days or at most a couple of weeks, until the defeated gang remobilized and reasserted its rights.

References

Adamson, C. 1998. 'Turf, Honor and the American Street Gang: Patterns of Continuity and Change Since 1820', *Theoretical Criminology* 2(1), 57–84.

Arendt, H. 1969. *On Violence*. New York: Harcourt Brace.

Arias, E. D. 2006. *Drugs & Democracy in Rio de Janeiro: Trafficking, Social Networks, & Public Security*. Chapel Hill, NC: University of North Carolina Press.

Banner, S. 2005. 'Why Terra Nullius? Anthropology and Property Law in Early Australia', *Law and History Review* 23(1), 95–131.

Bloch, M. 1996. 'La "Consommation" des Jeunes Hommes chez les Zafimaniry de Madagascar', in F. Héritier (ed.), *De la Violence*. Paris: Éditions Odile Jacob, 201–222.

Cruz, J. M. and A. Durán-Martínez. 2016. 'Hiding Violence to Deal with the State: Criminal Pacts in El Salvador and Medellin', *Journal of Peace Research* 53(2), 197–210.

Dimitriadis, G. 2006. 'The Situation Complex: Revisiting Frederic Thrasher's "The Gang: A Study of 1,313 Gangs in Chicago"', *Cultural Studies – Critical Methodologies* 6(3), 335–353.

Evans-Pritchard, E. E. 1940. *The Nuer: A Description of the Modes of Livelihood and Political Institutions of a Nilotic People*. Oxford: Oxford University Press.

Fischer, C. 1975. 'Toward a Subcultural Theory of Urbanism', *American Journal of Sociology* 80(6), 1319–1330.

Fitzmaurice, A. 2008. 'The Genealogy of *Terra Nullius*', *Australian Historical Studies* 38(129), 1–15.

Fontes, A. W. 2018. *Mortal Doubt: Transnational Gangs and Social Order in Guatemala City*. Berkeley, CA: University of California Press.

Geenen, K. 2009. '"Sleep Occupies No Space": The Use of Public Space by Street Gangs in Kinshasa', *Africa* 79(3), 347–368.

Glaser, C. 1998. 'Swines, Hazels and the Dirty Dozen: Masculinity, Territoriality and the Youth Gangs of Soweto, 1960–1976', *Journal of Southern African Studies* 24(4), 719–736.

Graham, S. and S. Marvin. 2001. *Splintering Urbanism: Networked Infrastructures, Technological Mobilities and the Urban Condition*. London: Routledge.

Gutiérrez Rivera, L. 2010. 'Discipline and Punish? Youth Gangs' Response to "Zero-Tolerance" Policies in Honduras', *Bulletin of Latin American Research* 29(4), 492–504.

Gutiérrez Rivera, L. 2013. *Territories of Violence: State, Marginal Youth, and Public Security in Honduras*. New York: Palgrave Macmillan.

Hazen, J. M. and D. Rodgers (eds). 2014. *Global Gangs: Street Violence Across the World*. Minneapolis, MN: University of Minnesota Press.

Hoelscher, K. and P. M. Norheim-Martinsen. 2014. 'Urban Violence and the Militarization of Security: Brazilian "Peacekeeping" in Rio de Janeiro and Port-au-Prince', *Small Wars & Insurgencies* 25(5), 957–975.

Jensen, S. 2008. *Gangs, Politics and Dignity in Cape Town*. Chicago, IL: University of Chicago Press.

Jones, G. A. and D. Rodgers. 2015. 'Gangs, Guns, and the City: Urban Policy in Dangerous Places', in C. Lemanski and C. Marx (eds) *The City in Urban Poverty*. London: Palgrave Macmillan, 205–226.

Jütersonke, O., R. Muggah, and D. Rodgers. 2009. 'Gangs, Urban Violence, and Security Interventions in Central America', *Security Dialogue* 40(4–5), 373–397.

Keiser, R. L. 1969. *The Vice Lords: Warriors of the Street*. New York: Holt, Rinehart, and Winston.

Kintrea, K., J. Bannister, and J. Pickering. 2010. 'Territoriality and Disadvantage Among Young People: An Exploratory Study of Six British Neighbourhoods', *Journal of Housing and the Built Environment* 25(4), 447–465.

Lepoutre, D. 1997. *Cœur de Banlieue: Codes, Rites et Langages*. Paris: Éditions Odile Jacob.

Manwaring, M. G. 2005. *Street Gangs: The New Urban Insurgency*. Carlisle, PA: US Army War College.

Mohammed, M. 2011. *La Formation des Bandes: Entre la Famille, l'École et la Rue*. Paris: Presses Universitaires de France.

Moore, J., D. Vigil, and R. Garcia. 1983. 'Residence and Territoriality in Chicano Gangs', *Social Problems* 31(2), 182–194.

Murray, M. 2017. *The Urbanism of Exception: The Dynamics of Global City Building in the Twenty-First Century*. Cambridge: Cambridge University Press.

Pickering, J., K. Kintrea, and J. Bannister. 2012. 'Invisible Walls and Visible Youth: Territoriality among Young People in British Cities', *Urban Studies* 49(5), 945–960.

Rocha, J.-L. 2007. *Lanzando Piedras, Fumando 'Piedras': Evolución de las Pandillas en Nicaragua 1997–2006*, Cuaderno de Investigación No. 23. Managua: UCA Publicaciones.

Rodgers, D. 2006. 'Living in the Shadow of Death: Gangs, Violence, and Social Order in Urban Nicaragua, 1996–2002', *Journal of Latin American Studies* 38(2), 267–292.

Rodgers, D. 2007a. 'Joining the Gang and Becoming a Broder: The Violence of Ethnography in Contemporary Nicaragua', *Bulletin of Latin American Research* 26(4), 444–461.

Rodgers, D. 2007b. '"Each to Their Own": Ethnographic Notes on the Economic Organisation of Poor Households in Urban Nicaragua', *Journal of Development Studies* 43(3), 391–419.

Rodgers, D. 2007c. 'When Vigilantes Turn Bad: Gangs, Violence, and Social Change in Urban Nicaragua', in D. Pratten and A. Sen (eds) *Global Vigilantes*. London: Hurst, 349–370.

Rodgers, D. 2008. 'Searching for the Time of Beautiful Madness: Of Ruins and Revolution in Post-Sandinista Nicaragua', in H. West and P. Raman (eds) *Enduring Socialism: Explorations of Revolution and Transformation, Restoration and Continuation*. Oxford: Berghahn, 77–102.

Rodgers, D. 2009. 'Slum Wars of the 21st Century: Gangs, Mano Dura, and the New Urban Geography of Conflict in Central America', *Development and Change* 40(5), 949–976.

Rodgers, D. 2010. 'Urban Violence Is Not (Necessarily) a Way of Life: Towards a Political Economy of Conflict in Cities', in J. Beall, B. Guha-Khasnobis, and R. Kanbur (eds) *Urbanization and Development: Multidisciplinary Perspectives*. Oxford: Oxford University Press, 235–248.

Rodgers, D. 2014. '*Bróderes*, *Vagos*, and *Compadres* in the *Barrio*: Kinship, Politics, and Local Territorialization in Urban Nicaragua', in B. Fischer, B. McCann, and J. Auyero (eds) *Cities from Scratch: Poverty and Informality in Urban Latin America*. Durham, NC: Duke University Press, 127–149.

Rodgers, D. 2015. 'The Moral Economy of Murder: Violence, Death, and Social Order in Gangland Nicaragua', in J. Auyero, P. Bourgois, and N. Scheper-Hughes (eds) *Violence at the Urban Margins*. Oxford: Oxford University Press, 21–40.

Rodgers, D. 2017. '*Bróderes* in Arms: Gangs and the Socialization of Violence in Nicaragua', *Journal of Peace Research* 54(5), 648–660.

Rodgers, D. Forthcoming 2019. 'From "Broder" to "Don": Methodological Reflections on Longitudinal Gang Research in Nicaragua', in K. Koonings, D. Kruijt, and D. Rodgers (eds) *Risky Anthropology*. Lanham, MD: Lexington Books.

Salagaev, A. L. and R. R. Safin. 2014. 'Capitalizing on Change: Gangs, Ideology and the Transition to a Liberal Economy in the Russian Federation', in J. M. Hazen and D.

Rodgers (eds) *Global Gangs: Street Violence Across the World*. Minneapolis, MN: University of Minnesota Press, 65–84.

Saunders-Hastings, K. 2015. 'Order and Insecurity under the Mara: Violence, Coping, and Community in Guatemala City', unpublished PhD, Centre for Socio-Legal Studies, University of Oxford.

Sen, A. 2014. '"For Your Safety": New Cultures of Gang Formations in Urban Indian Slums', in J. M. Hazen and D. Rodgers (eds) *Global Gangs: Street Violence Across the World*. Minneapolis, MN: University of Minnesota Press, 193–212.

Stephenson, S. 2015. *Gangs of Russia: From the Streets to the Corridors of Power*. Ithaca, NY: Cornell University Press.

Suttles, G. D. 1968. *The Social Order of the Slum: Ethnicity and Territory in the Inner City*. Chicago, IL: University of Chicago Press.

Thrasher, F. 1936 [1927]. *The Gang: A Study of 1,313 Gangs in Chicago*. Chicago, IL: University of Chicago Press.

Tita, G. E., J. Cohen, and J. Engberg. 2005. 'An Ecological Study of the Location of Gang "Set Space"', *Social Problems* 52(2), 272–299.

van der Borgh, C. and W. Savenije. 2016. 'The Truce Between the Mara Salvatrucha and 18th Street Gang in El Salvador: From Opportunity to Failure', unpublished working paper, available at: www.researchgate.net/publication/309397890_The_Truce_between_the_Mara_Salvatrucha_and_18th_Street_Gang_In_El_Salvador_From_Opportunity_to_Failure (accessed 1 July 2018).

Whyte, W. F. 1993 [1943]. *Street Corner Society: The Structure of an Italian Slum*, 4th edn. Chicago, IL: University of Chicago Press.

Wolf, S. 2017. *Mano Dura: The Politics of Gang Control in El Salvador*. Austin, TX: University of Texas Press.

Wolseth, J. 2011. *Jesus and the Gang: Youth Violence and Christianity in Urban Honduras*. Tucson, AZ: University of Arizona Press.

4 Negotiating with criminal armed groups

From prejudice to pragmatism

Hugo van den Eertwegh

Introduction

Criminal armed violence has become an important challenge for many cities and neighbourhoods. The flexible, networked, and global nature of many criminal armed groups (CAGs), their access to financial and human resources, and their extreme cruelty and absence of any ideology have created difficulties for strategists and policymakers alike in developing efficient responses (Perez-Salazar 2016; Briscoe 2013). The problem of finding adequate policy responses to criminal armed violence is often a major issue in large cities that are already grappling with uncontrolled urbanization, ungoverned spaces, and a complex symbiosis between urban neighbourhoods and rural hinterlands. Coupled with corruption and impunity at different levels of society and government, this sociopolitical landscape has become a powerful playing field for CAGs (Muggah 2015; Briscoe *et al.* 2014; Schultz *et al.* 2004).

In recent years there has been more interest in preventive or holistic approaches to armed or criminal violence, as exemplified by the 100 promising practices on safer cities (Eguizabal *et al.* 2014). Such responses exist frequently in contrast (and opposition) to securitized policies that promise to tackle criminal armed violence though a security prism. The application of securitized strategies has resulted in regular cycles of hard-line policies that have often increased rather than reduced levels of violence (Wolff 2015; IACHR 2015).

This chapter seeks to open the discussion around whether or not negotiating with CAGs could be an option to reduce levels of violence in the long term, arguing in favour of integrating this option into the violence reduction toolbox. This argument is rooted in the recognition that armed violence is a major threat to many societies, and solutions for the problem require recourse to the full range of policy options, including negotiation strategies with perpetrators of violence and building the technical skills to conduct negotiations with CAGs when opportunities exist. The chapter looks at the prejudices underlying negotiations with CAGs, and at ‘how’ and ‘when’ it is advisable to apply negotiation as a strategy to reduce and prevent violence. The chapter argues for more innovative and inclusive forms of citizen participation, both to strengthen the democratic foundations of negotiations with CAGs and to develop more sustainable negotiation outcomes.

This chapter understands armed violence as 'the intentional use of illegitimate force (actual or threatened) with arms or explosives, against a person, group, community, or state that undermines people-centred security and/or sustainable development' (Geneva Declaration Secretariat 2008). It also uses the generic term of 'criminal armed groups' to cover all types of non-ideological groups actively engaged in armed violence to sustain criminal agendas, including a wide range of actors from street gangs to cartels. CAGs can be considered as part of the generic group of 'non-state armed groups' (NSAGs). NSAGs still lack an official definition, thus for the purposes of this chapter I define them as armed groups not under the control or command of one or more states, and which are party to a situation of violence of concern. This definition does not enter into a determination of whether or not a group is an 'armed group' as defined under international humanitarian law; this will depend on levels of violence generated and internal organization. The chapter illustrates its argument mainly with examples from Latin America.

Assessing the risks to negotiate

When discussing negotiations with armed groups one usually encounters a range of staunch objections which turn around concepts of sovereignty and justice, and are peppered with flakes of morality. The latter has been underlined by the rejection of the first version of the Colombian peace agreement, where the concept of granting political rights to the Fuerzas Armadas Revolucionarias de Colombia – Ejercito del Pueblo (FARC–EP) leadership and transitional justice concessions proved an easy lever for opposition forces to use and helped ensure victory in a referendum (Vivanco 2016). Colombia is not alone when it comes to popular support for state repression of crime groups. In El Salvador the 2012–2014 collapsed truce process testified to the enduring tension around the subject of interaction with armed groups, especially CAGs (Hernandéz-Anzora 2016). Despite their popularity among middle and upper classes, hard-line policies have so far failed to produce any significant reductions in levels of crime and violence. On the contrary, in many contexts they reduce the security of citizens and service delivery, and increase homicide rates, the number of missing persons, and internal or interurban displacement (Kinosian *et al.* 2016; Cantor 2014).

The first part of this chapter discusses three types of risks generally related to negotiating with CAGs, and explores why they provide insufficient grounds to dismiss negotiations with CAGs as a tool for ending chronic levels of urban violence.

Risk 1: entering into negotiations with CAGs will lend (political) legitimacy to them

The concern of legitimizing armed groups is an old fear of state and city authorities. It harks back to the Weberian political concept that legitimacy rests on popular acceptance of an authority (Bensman 2015). As such, legitimacy is part of the social contract between the state and its citizens in which consent, in exchange for security and services, has largely replaced state coercion. To be able to protect the security of its population the state has a monopoly on the use of force. In many countries, through their actions – both violent and non-violent – CAGs challenge this monopoly on use of force and state service provision, undercutting the precepts of the social contract and state legitimacy.

One could therefore argue that once a state is unable to establish control of a certain area of territory within its borders – or a specific part of a city – a de facto transfer of legitimacy takes place to the controlling group through 'domination' of that area. This controlling group in many contexts can be characterized as CAGs. In some contexts the domination may fluctuate constantly between opposing groups, but the absence of the state in providing basic services and, especially, security for its citizens often remains a constant. Zaluar describes, for example, the situation in Rio de Janeiro's *favelas*:

> Drug Commandos have transient skirmishes to dispute territory within the favelas where their markets are located. As a result of the ensuing military-type control, in most areas inside favelas or near them, the drug lords or 'donos' [owners] restrict the movements of residents and government agents, therefore limiting access to public services such as schools, health agencies and sports compounds. Residents of one favela cannot enter the territory of an 'enemy' favela, even when delivering goods, visiting friends and relatives, or when on dates. If they do, they are killed, especially if they are young men.
>
> (Zaluar 2010: 17–18)

An even more explicit statement was made by former Mexican President Felipe Calderón, who said that cartels are

> … imposing fees like taxes in towns they dominate, extorting money.… This has become an activity that defies the government, and even seeks to replace the government … trying to impose a monopoly by force of arms … even trying to impose their own laws.
>
> (AP 2010)

In reaching this level of domination, the *modi operandi* of CAGs and NSAGs converge in terms of the illegality of both their methods to challenge the state and its monopoly of violence and their means to carry out their struggle, as financing is related to criminal activities. To achieve their goals, both types of armed group resort to, among many other activities, assassinations, kidnapping, corrupting state officials, confronting state security forces, forced recruitment, illegitimate or legitimate business, and setting up alternative tax systems in the shape of 'voluntary' revolutionary taxes or extortion (ICG 2017; Rexton Kan 2015). What differs is that the end goal for NSAGs is to replace or change a regime or achieve a specific political agenda, whereas for CAGs it suffices to 'shape' or 'restructure' governance to fit the purpose of criminal agendas. In many contexts this means hollowing out state authority at a local or national level, without necessarily replacing the 'shell' or 'appearance' of such authority (Wennmann 2014; Tager and Umaña 2013; Hagedorn 2005; Monzón 2017). By getting into the business of dominating territory, both CAGs and NSAGs move into the legitimacy realm of the state through the provision of security and services – activities reserved for the state under a Weberian notion of governance (Reitano and Hunter 2016).

In this context, the question is what more legitimacy can state authorities lose through considering negotiations with CAGs? Some observers point out that CAGs gain a stage to present their claims at local, national, or international levels, garnering both additional internal and external support as well as further strengthening their credibility in their communities (van der Borgh and Savenije 2015). Even though these risks exist, past experiences in peace processes with groups like the Liberation Tigers of Tamil Eelam, the Irish Republican Army (IRA), and Euskadi Ta Akatasuna (ETA) have shown that any additional perceived legitimacy due to participation in official peace negotiations did not survive a failed process. Any obtained legitimacy was only preserved if groups actively contributed to the implementation of peace agreements (Petrasek 2004a, 2004b; Powell 2014; Hernandéz-Anzora 2016). Also, it is questionable whether in the current age CAGs or NSAGs still need such a 'platform' to enhance their legitimacy.

A final observation is that CAGs usually do not reach the levels of control and domination described above without substantial 'assistance' from the state itself, through corruption, impunity, and poor governance (Hernandez 2012; Dickie 2013; Briscoe 2015). This raises the delicate point that 'the state' and 'CAGs' are usually not two distinct and opposed sets of actors, but are often connected to each other in various ways and at various levels (Perdomo and Burcher 2016). The question of what lends legitimacy to CAGs therefore becomes more complicated. For instance, excessive violence by state or city police forces against resident populations risks contributing to strengthening local legitimacy of CAGs in the eyes of the people suffering these abuses. In this regard, a study on Central America shows that under 'some circumstances police transgressions have a greater impact on the legitimacy of the political system than crime or insecurity' (Cruz 2015: 251; see also Piché 2016).

Risk 2: negotiations will simply give CAGs time to (re)organize themselves

CAGs abusing negotiations to win time or strengthen their overall criminal structure is certainly a potential risk, but it applies across the board to negotiations with any armed group (Cockayne 2013). As Zartman (2001: 9) notes: 'It is difficult at the outset to determine whether negotiations are indeed serious or sincere, and indeed true and false motives may be indistinguishably mixed in the minds of the actors themselves at the beginning.'

In El Salvador the argument that gangs abused negotiations to strengthen their structures has been strongly put forward in opposition to the truce process between rival *maras* (Lohmuller 2015; Dudley 2010). As an argument against any negotiation, this is fair enough. However, if opponents are really so interested in preventing the CAGs being strengthened with government complicity, they should not shy away from looking closely at failed (security) policies over the last decade and their impact on strengthening CAGs in the first place. Several studies show that it was the different *mano dura* policies in the early 2000s that pushed *maras* in El Salvador to reorganize, regroup, and recruit (ICG 2017; Hazen and Rodgers 2014).

Under these 'tough-on-crime' policies, the government introduced anti-gang laws in 2003 that criminalized gang membership and penalized it with up to five years in prison (ALRES 2003: Art. 6; Umaña *et al.* 2013). These policies were reinforced by another round of anti-gang laws in 2009 (Piché 2016). In 2014, once the truce process had fully collapsed, the Salvadorian government again harked back to heavy-handed policies, including officially declaring the *maras* a terrorist group. Simply belonging to a *mara* was enough to incur prison sentences of up to 15 years (Kinosian *et al.* 2016; Daugherty 2015). 'Tough-on-crime' policies were compounded by assassinations of gang members by security services (Ungar 2009; ICG 2017). Heavy-handed government policies targeted entire groups of youth, deepened their sense of exclusion, and bolstered the defiance identity of gang members (Cruz 2011; Zaluar 2000, 2004). This process is described by 'El Viejo Lin', national leader of the Barrio 18 gang in El Salvador:

> The persecution of our brothers has been the fertilizer that has made the gangs grow. They capture five this week. Within a month, there are ten more. They put the father in jail, his kids are next. When those kids grow up, then the grandkid is next. He has a wife, she gets pregnant, that kid is already part of it. Do they achieve anything massacring or jailing us? We doubt it....
>
> (Quoted in Dudley 2013)

As an example of the arbitrariness of 'the anti-*mara* laws', 19,275 persons were arrested between July 2003 and August 2004 but only 975 eventually received a sentence (ICG 2017).

In response, the *maras* organized to defend themselves, and imprisonment helped consolidate group ties and strengthen hierarchy (Muggah and De Carvalho 2014). Developing links with transnational organized crime provided income and weapons, intensifying clashes between gangs and pushing El Salvador into a downward spiral of escalating violence (Kinosian *et al.* 2016; Cantor 2014). This trend came on top of inherent state fragility due to the preceding civil war and deportations from the United States of many thousands of youths, many with criminal records (World Bank 2011).

The El Salvador example underlines a key point: the risk that negotiations are used by CAGs to strengthen their structures needs to be placed into the wider policy context of heavy-handed measures against CAGs and their effects on their development. It also needs to be recognized that abusing a negotiation process to strengthen operational capacity is of course not only the prerogative of CAGs. Government forces play the same game, and also use negotiations to buy time and catch opponents off guard to deliver a military blow. These are common strategic dilemmas of negotiation processes, and well recognized in the literature on the role of violence during peace processes (Höglund 2008).

Risk 3: negotiating with CAGs undermines the rule of law, disrespects victims, and incentivizes people to enter into crime

To fight crime, many governments make significant inroads into the rule of law in their countries. This is especially the case when emergency criminal justice regimes erode constitutional rights and democratic values in the struggle to defeat CAGs (IACHR 2013, 2017; UNHRC 2013). Anti-crime laws are a good example. In Mexico the *arraigo* concept (detention without charge) allows the arrest and preventive detention for up to 80 days of anyone suspected of involvement in organized crime (Deanton and Ferreira 2015: 31–32; Escobar 2014). Another example is the Colombian emergency legislation that has for decades created a de facto parallel criminal justice structure in which citizens' rights are significantly reduced (Lajous 2014). In addition to these legislative measures, there is regionwide use of the military in law enforcement, with tens of thousands of troops carrying out public security functions, often without the proper training and equipment or coordination between forces. Neither is there much oversight, nor clear policies on when the deployments might end. A recent decree from Brazilian President Temer again ordered the military into the streets of Rio de Janeiro to fight the gangs and regain control of parts of the city, making it the 29th military operation in the city since 2001 (Phillips 2018).

In addition, impunity reigns in regions like Mexico and Central America. Very few crimes are investigated, and even fewer investigations actually lead to convictions (Open Society Foundations 2016). The Northern Triangle between Guatemala, Honduras, and El Salvador is an example for the region. Between 2011 and 2013, 48,947 people were murdered but in only 2,295 cases were perpetrators actually convicted, representing an impunity rate of some 95 per cent (Olson 2015; Le Clerq and Lara 2015). On top of this, victims reaching out to the state often find themselves stigmatized and dealing with institutions ill-equipped to meet their needs.

In the context of impunity, negotiations with CAGs can put further pressure on the perception of justice. This is because negotiations imply the acceptance of concessions with CAGs in return for less violence. Such concession making is not new, and no less painful than transitional justice arrangements after interstate or intrastate armed conflicts or dictatorships, as illustrated by the Colombian peace process and the peace agreement between the Afghan government and the Hezb-e-Islami (Vivanco 2016; Gossman 2016). Both include extensive transitional justice concessions, in the Afghan case even including immunity from past war crimes. Most of these concessions do not have general public approval, nor do they respect the memories of the thousands of victims of these groups over the decades. Similarly, deals done within the normal criminal justice system do not necessarily provide satisfactory results for individual victims (Gilna 2014; Barton 2012).

The question is thus rather which instrument is most appropriate in a certain circumstance to achieve the greater public good and find a politically and publicly acceptable balance between different tensions within a society. The Truth and Reconciliation Commission (TRC) in South Africa, for example, together with the transition to a government headed by the African National Congress, have mitigated the absence of criminal prosecution of members of the apartheid regime. Other trade-offs were made to reach agreements with ETA and the IRA. None of these solutions was perfect or satisfactory to all, and for former South African President Frederik Willem de Klerk the absence of actors such as the National Party and the Inkata movement from the TRC left South Africa more divided then ever when the TRC ended (Powell 2014). But regardless of these imperfections, the TRC was considered a politically acceptable solution by the actors driving South Africa's transition at the time.

Given the challenge of violence, crime, and injustice in Latin America, these examples of transitional and restorative justice arrangements raise the question of whether they can be applied as tools to deal with situations of criminal violence. If there is the space in major political transitions or after civil wars to come to an agreement with NSAGs, perpetrators of war crimes and crimes against humanity, and 'terrorist groups' or members of dictatorial regimes, there should be the political space to work towards arrangements between CAGs, state actors, and other concerned parties to reduce criminal violence.

Overall, high levels of corruption and impunity should weigh considerably in the balance on what real justice and respect for the rule of law for the population is all about (Armao 2014; Farah and Meacham 2015). This of course also links to the question of what 'incentivizing crime' means in the first place. Rather than possible negotiations with CAGs, it is more often the state itself through corruption and lack of opportunities which gives the signal that 'crime pays' and there is really no other option to make a decent living.

Precedents of negotiations with criminal armed groups

'Politics and the Mafia are two forces that share the same space: they either fight each other or come to an agreement' (Abbate and Gomez 2007: 37). This statement by Italian anti-Mafia magistrate Paolo Borselino (who was assassinated by the Italian Mafia in Palermo in 1993) underlines the political reality that states either fight or come to an 'agreement' (hence they negotiate) with CAGs. In the latter case, this is usually done regardless of the risks previously mentioned. This section reviews several examples of arrangements between state actors and CAGs to demystify the issue of negotiations with CAGs.

To start with, there are negotiations within the criminal legal framework. These negotiations are common, and mostly concern arrangements between a specific individual who has committed a crime and law enforcement agencies. Usually such arrangements (also known as plea bargaining) involve the offer of a reduced sentence or protection in exchange for information on a criminal network or high-level crime leaders. In Italy there are the *pentititi*, former members of the Mafia who have turned state's evidence and have been important in giving Italian law enforcement agencies insights into the workings of the different organized crime structures in Italy, such as the Cosa Nostra (Dickie 2013; Bodrero 2013). In El Salvador the construct of the *testigo criteriado* has been used widely to plea bargain with incarcerated gang members (UNHCR 2013). These legitimate negotiations remain within the confines of the criminal justice system, with no or very limited involvement of victims, and often remain veiled in secrecy, especially when they concern high-profile cases (Carpenter 2016). Such deals, apart from showing little respect for victims, risk huge unintended consequences. This was shown by the deal made with the former boss of the Gulf Cartel, Osiel Cardenas Guillen, in 2010, details of which have only recently started to come out. Cardenas, in exchange for a relatively lenient 25-year sentence (not in a supermax facility), gave US authorities some US$50 million in cash as well as a wealth of information on his organization, including the Los Zetas group. This resulted in an extremely violent 'war' between the Zetas and the Gulf Cartel that has claimed hundreds of lives (Corchado and Krause 2016).

Another form of negotiation with criminal groups relates to the use of informants from within the groups, infiltrated agents, and different types of sting operations. Such approaches open up grey areas of policymaking where political and public interests are constantly balanced, allowing lower-level criminal activity to take place so as to crack down on higher echelons of a criminal network. These are all engagements that include elements of negotiation, be it direct (between state and informant) or indirect (allowing certain activities to continue), and they usually occur in secrecy, with limited oversight, and at times help to strengthen CAGs. An example of the latter is the 'fast and furious' affair. Here the US Bureau for Alcohol, Tobacco, Firearms and Explosives allowed the 'walking' of weapons from US arms dealers via straw purchasers across the border into Mexico so as to arrest the buyers. In the process some 2,000 weapons, many of military grade (AK-47, AR-15), 'walked' into the hands of Mexican drug cartels. Apart from the fact that US agencies lost track of more than half these weapons, many were later identified at crime scenes in Mexico. When the affair was exposed it led to political tensions between the United States and Mexico, as well as within the US government (US Department of Justice 2012). This raises the question of how risks were evaluated (and against what) when taking the decision to go ahead with such an operation, compared to the risks which are quickly presented when proposing negotiations with CAGs.

Related to the above are covert operations that illustrate how government entities and CAGs interact and negotiate to attain some broader objective. One example is the weapons deliveries by the US Central Intelligence Agency to the *contras* in Nicaragua in exchange for allowing drugs to enter the United States (*United States of America* v. *José Rafael Abello Silva* 1990; Gagne 2015). A well-documented example from Turkey has been the use by authorities of criminals to fight the Kurdistan Workers' Party and left-wing opponents. Those working with the authorities were subsequently shielded from prosecution and given considerable protection to carry out their illegal business (Massicard 2010).

Another type of arrangement is a tacit agreement between a government and criminal groups to respect each other's space. This allows the government to present a calm and relatively peaceful environment, while at the same time the criminal groups can carry out their business without much interference from law enforcement. A classic example is the *entente cordiale* between different Mexican governments and a range of different criminal groups (Grillo 2012; Santamaria 2014). A common feature of such arrangements in Central and South America is the financing of candidates in (local) electoral campaigns by CAGs to obtain political support, including influencing the drafting of legislation (Perdomo and Burcher 2016; Reitano and Hunter 2016; ICG 2017). Another example is the negotiations between the Italian government and the Sicilian Mafia (Cosa Nostra). This occurred in the early 1990s after a spree of bombings and killings by the Mafia aimed at a cessation of the violence in exchange for serving prison time in less harsh prison regimes (Torrealta and Mottola 2012). A similar negotiation took place in Brazil in 2006 after the Primeiro Comando da Capital unleashed a wave of violence in São Paulo (Maloione and Malone-Rowe 2014). In addition there is a wide array of commercial agreements, ranging from triangulations between (indigenous) land rights to complex money-laundering schemes (Briscoe *et al.* 2014; Paley 2014).

Another type of practice is humanitarian negotiations with criminal groups, mainly around specific issues such as access for health personnel, safe access to schools, or psychosocial support for victims of violence in violence-affected regions or neighbourhoods (Lucchi 2014). All these activities take place in areas where CAGs are present or in full control, and imply at least tacit agreement on their side (and on the side of authorities) to the services provided by humanitarian actors; this is achieved by some level of indirect negotiation, often through the communities (Bangerter 2010). An example of more direct engagement is the work done by so-called 'interrupters' used by Cure Violence. Interrupters have the delicate task of preventing retaliatory violence after an incident has happened, and as such interact directly with CAG members on not seeking revenge (Cure Violence 2016; see also the discussion by Austin and Jütersonke in Chapter 1 of this volume). Other examples are local non-governmental organizations involved in community-level negotiations and religious bodies or churches that work in violence-affected neighbourhoods (IDMC 2015).

A last example of negotiations with CAGs is gang truces, of which there have been many over the last 30 years in a wide variety of countries, such as the United States, Brazil, Colombia, Belize, Honduras, Trinidad and Tobago, Jamaica, and El Salvador (Cruz and Duran-Martínez 2016; Planta and Dudouet 2015). Gang truces in Belize and El Salvador had a positive start in terms of their ability to reduce violence, but they could not be maintained. The reasons are manifold, and include for instance the gangs' dishonesty but also – in the case of El Salvador – intervention of powerful spoilers within the political establishment who torpedoed the truce and a lack of popular support linked to the upper middle class (a segment of society that is not necessarily the most affected by armed violence) (Bustelo 2015). With regard to the Belize gang truce, money and commitment ran out (McDermott 2012). Colombia has over time also experimented extensively with truces or negotiations with CAGs. At the forefront has been Medellín, a city plagued by violence, where in the 1990s there were mediation efforts between communities and local government with some 160 different CAGs (Velez 2001). In 2003 as part of the demobilization negotiations between the Colombian government and the paramilitary groups (Autodefensas Unidas de Colombia), negotiations with the Oficina de Envigado (which was integrated into the Metro Block self-defence group) that largely controlled criminal activity in the city managed to achieve a significant drop in homicides for several years. This agreement, however, was fully linked to the main CAG strongman in Medellín (Diego Fernando Murillo, also known as Don Berna), and when he was extradited to the United States on drug charges in 2008 the truce collapsed. Further truces between factions were agreed in 2008 and 2013.

In sum, this section suggests that there is a significant track record of interaction and negotiations of different types of state agencies with CAGs, and that such interaction and negotiation are not uncommon. The section points to the practices of plea bargaining, engagement of informers, kidnapping negotiations, covert operations, humanitarian negotiations, and gang truces as well-established approaches used by different state agencies to interact with CAGs. Yet in many contexts such interaction is illegal, hidden, and ad hoc, and it occurs with little or no (democratic) oversight (Martínez d'Aubuisson and Martínez 2016). Overall, this track record of practice calls into question the validity of the mantra that 'states do not negotiate with criminals'. If this mantra does not reflect the actual practice, the question arises of why not transform the issue of negotiations with CAGs into policies that are more constructive and transparent, and at the same time more effective in seizing opportunities which offer the chance to reduce armed violence and improve urban safety. To explore how to move towards such policies, the next section analyses criteria for engaging with armed groups and reflects on 'when' and 'how' to engage CAGs.

Criteria for engaging CAGs

The study of criteria for engagement with CAGs starts from the recognition of three elements that emerge from the preceding analysis. First, negotiating with CAGs implies risks, but this is similar to negotiations with other armed groups and hence should be approached the same way. Second, the behaviour and capabilities of governments and other state actors play an important role in shaping CAGs and the violence they produce. Heavy-handed security policies, high levels of impunity, and endemic corruption affect the nature, strength, and activities of CAGs. This should be added to the equation when assessing risks of negotiating with CAGs. Engagement is not one-sided, but includes both state and non-state dynamics of organized crime. Third, there is a substantial practical record of interaction between different state authorities and CAGs in many contexts; yet most of such practice does not focus on holistic violence reduction and is usually ad hoc, focused on homicide reduction, and lacks oversight.

In many situations ending criminal violence through negotiations is a concrete opportunity and can be a policy alternative to bolster safety and security, especially in urban contexts. The attention on the issue of negotiations with CAGs should therefore focus on 'when' and 'how' to negotiate, rather than 'if' or 'why'.

When to engage

The concept of the 'ripeness' of a process is a useful heuristic tool to understand moments and opportunities for engagement. Ripeness describes a situation where conflict parties have reached a 'mutually hurting stalemate' (MHS), which occurs 'when parties find themselves locked in a conflict from which they cannot escalate to victory and this deadlock is painful for both of them' (Zartman 2001: 8).

How can the MHS concept be applied to CAGs and situations of armed violence? CAGs are not seeking a particular victory, so when do they reach their version of the MHS? What may be different with CAGs in comparison to NSAGs is that the concept of 'victory' is more individual. Victory for many individuals involved in organized crime is defined more by the continuation of a certain lifestyle than by the achievement of an overarching strategic goal. With regards to CAGs, the concept of 'stalemate' – a position or situation in which no action can be taken or progress made – can be understood as a moment when individual actors or a majority of CAG members reach a point of disillusionment around the realization that they are stuck in perpetual cycles of violence from which the only escape is jail or death. CAG members are usually aware of this fact, and often even brag about it as a part of their way of life (*la vida loca*). Older gang members in particular can come to a point when they develop a more nuanced understanding of gang life, and alternatives to crime and violence become attractive (Martínez and Valencia 2017).

These dynamics were expressed clearly by the leaders of the different *maras* during the truce process in El Salvador. The leaders were looking for a way out of violent gang life (Mijango 2013). Zaluar describes a similar sentiment when talking with gang members released from prison in Rio de Janeiro:

> When they talk about the past, they do not seem to miss it or wish to go back to the same path. Although some liked to boast of their fearless actions confronting enemies, displaying prowess and pride, most referred to this time as a very difficult one because suspicion, self-interest and the yearning for money made real friendship impossible. The images they report are of a time in which they were surrounded by enemies and had to watch their

> backs constantly, something I also registered when interviewing young boys on the verge of becoming traffickers.
>
> (Zaluar 2010: 22–23)

None of the persons interviewed had become rich, as everything had disappeared in 'corrupt policemen, lawyers, women, loans to family members, relatives or friends, parties and feasts' (ibid.). After release from prison the gang members were ready to move away from crime, but found it impossible to get jobs. One reason was that Brazilian law obliges job applicants to produce a document certifying that the applicant has never been arrested (ibid.).

These reflections about MHSs are relevant for the engagement of and negotiations with CAGs. If, as in El Salvador, the existence of an MHS is expressed by a gang's leadership, there is potential for them to transpose this view on to a larger group simply due to their degree of hierarchical control over the peer network in which gang members interact. Gang leaders have an impact on behaviour well beyond any other factor, including family (Hagedorn 2005; Zaluar 2010). This is reflected by statements of Barrio 18 leader El Viejo Lin when asked what younger gang members thought of the truce process:

> For the younger generation, this doesn't matter. They are there because we have sat them down, explained it to them. We are on top of them. And they're getting it, they're getting it. This is not easy. If this doesn't work, we don't have any idea what will happen.
>
> (Quoted in Dudley 2013)

On the side of governments, the issue of victory over CAGs is more straightforward; it often means annihilating a CAG. Thus a stalemate comes about when government actors feel the cost of continuing to fight on without any prospect of reducing crime or violence. The prejudice against CAGs and a government's 'moral obligation' to fight crime tend to erode its capacity to accept that an MHS has been reached. This can be compounded by electoral cycles where candidates have to show their toughness on crime to gain votes.

The concept of 'ripeness' is about the antagonistic relationship between 'warring parties' to a conflict reaching an MHS (Zartman 2001). To make sense of ripeness with respect to criminal armed violence, it is necessary to expand our understanding of who counts as a party to a conflict. In the case of criminal armed violence, a central and often neglected party is the public, especially the population that is living in violence-affected neighbourhoods or regions, because they have the biggest stake in the reduction of violence.

This is especially the case when corruption, impunity, and poor governance make state authorities a critical part of the violence dynamics, so they do not truly represent the people affected by violence (Koonings and Kruijt 2007). Once the separation of powers – a feature that lies at the heart of democratic states – becomes eroded due to endemic levels of corruption and impunity, the state becomes (at least partially) an actor on the problem side, and to a certain extent loses its democratic prerogative as a representative of its citizens. These characteristics reduce the legitimacy of the state to act as a spokesperson for the public. In consequence, there is a vacuum in the public policy space to advocate credibly for violence reduction (ibid.). This vacuum can be filled by civil society representing different non-state or non-CAG interests, and as such becoming a separate actor in a violence reduction processes and associated negotiations.

With respect to an MHS, when considering 'the public' as a separate actor it is possible to locate an additional dimension of stalemate that in turn can foster the acceptability of negotiations with CAGs and the sustainability of a broader transformation process. An MHS for the public could be quite different than for a government, because the issue of insecurity is often linked to both CAGs and state security forces and police. Adding the public or their representative civil society organizations as actors to negotiation processes implies a more diverse negotiation space. It opens the sensitive issue of the degree to which the government or different state authorities represent the views and interests of all segments of society, especially those most affected by violence and crime. It also opens the practical challenge of 'how' the participation of the public as an independent actor should occur in a negotiation process.

A last element with regard to the 'ripeness' of an environment to start negotiations is the issue of timing. To start any process at the end of a political cycle, for instance, would be inadvisable. Negotiating with CAGs is simply too easy a target for any political opposition to attack. A better use of pre-election periods is to prepare the ground, so that after elections a process can start. Part of assessing whether 'timing' is right is the willingness of parties to work towards forward-looking solutions, i.e. to focus on negotiations to build a better future rather than on the harm done in the past and retributions for this (Zartman 2005). This links to the argument that negotiations with criminal groups should not be a stand-alone violence reduction effort, but part of a larger shared vision to rebuild a society. Thus when considering timing one also needs to look at whether parties (including the public) have a shared perception of the situation, as this will greatly facilitate parties' willingness to come to a 'rich' compromise, as Haas (1988) terms it. Timing also concerns socioeconomic developments and geopolitical realities, which all need to converge at the right time.

The abovementioned elements involved in the ripe time for negotiations to start are of course only that. They are possible conditions that could help nurture a positive outcome prior to negotiations; they are not hard criteria. In addition, the fact that an environment is not yet fully ready could mean that third-party actors have a more active (preparatory) role to play to enable parties to move towards a more fertile negotiation space (Yawanarajah 2016).

How to engage

Negotiation practice underlines that clarity about a desired end-state for a negotiation and careful sequencing in the shift away from criminal rents are important when engaging CAGs. Replacing rents with other benefits can be especially tricky for CAGs because, unlike NSAGs with a political agenda, their motivation is linked to making profits. Criminal agendas risk leading to a zero-sum game, because CAGs will find little appeal giving up the proceeds from criminal activity. At the same time, governments have little political space for allowing CAGs to continue criminal activities in an interim period because such concessions are unpopular with the electorate in many contexts (Cockayne 2013).

Transiting from violent to more peaceful societies implies replacing shadow or coping economies, and this requires time and the acceptance that change does not come in an ideal sequential process. Markets need to be able to absorb any additional labour, and policymakers need to understand the impact of shadow economies on households and communities (Hagedorn 2005; Ballentine and Nitzschke 2005). Transitions from criminal violence also mean accepting a degree of coexistence between social groups that are perceived as legal or illegal by different parts of the population (UNODC 2012). Such processes are inevitably sensitive and vulnerable to spoiling, and to succeed need strong commitment on all sides, a willingness to understand the workings of CAGs, and a lot of patience and trust by all involved. Social reintegration of individuals who in their lifetimes have seen only violence and exclusion is a daunting task in most societies, and it means fighting prejudices on a daily basis and ensuring reforms of justice and political systems.

What is offered to a CAG during a transition period? Will a population accept such offers? What will victims receive as compensation? An underlying dilemma that frames responses to these questions is how 'to mobilize and sustain the economic transformation required to resolve criminal agendas without losing social support for the process' (Cockayne 2013: 10–11). This dilemma is especially pronounced when a process is in crisis and the 'balance of payment' between disbursed funds and the reduction in violence and crime is perceived as uneven by a significant portion of the electorate or key stakeholders in a transition process.

Public participation – or rather having 'the public' as a distinct actor in a negotiation – could be a possible strategy to reduce the impact of this dilemma of losing social support for a transition process. Involving stakeholder groups representing broader societal interest is nothing necessarily new from a peace process perspective: 'inclusive participation' in peace processes is now a common prerequisite in negotiations to end interstate or intrastate armed conflicts (Paffenholz 2014). However, as a party to a negotiation, civil society could directly participate in and shape discussions, rather than merely being a bystander. So far public participation in peace processes has never really been at par with the other parties, although negotiations in, for instance, Colombia, Guatemala, and Mali were quite advanced with regard to the level of direct public involvement in the peace process (Alvarez and Prado 2002; Schirch 2009; Algar-Faria 2013; Susskind 2010). Constituting the public as an actor in a transition process may create a triangular negotiation environment that could shift a negotiation process beyond a bilateral framing, and situate a government's partisan interests more clearly with respect to the interests of different segments of society (Podder 2012). While for some actors constituting the public as an independent actor in transition processes might be considered as going against the sovereign prerogatives of a state, other actors may emphasize that public participation in a transition process sets the foundations for redemocratizing politics, which for too long has been secretive, exclusive, and elite-centred.

The Colombian peace process exemplifies how a complex negotiation was technically very well handled but still did not manage to garner broad public support from Colombian society. After the rejection of the peace agreement in a referendum and subsequent renegotiations, the second version of the peace agreement was not presented to the public in a referendum but run through the more manageable gauntlet of Congress and the Senate, where it was ultimately adopted. Though politically understandable in view of the rejection in the earlier referendum, the choice of this simpler political trajectory has weakened the democratic foundations of the agreement and potentially endangered its future implementation (Idler 2016; *The Economist* 2016).

Including the public as an actor in a negotiation process requires substantial preparation, in both the selection of participants and their technical briefing, as well as possibly in pre-negotiations with CAGs. However, it allows a much more bottom-up and transparent process in which the items on the negotiation agenda could be informed by the priorities as identified by populations of neighbourhoods prone to crime and violence, including issues such as state repression, corruption, and impunity.

How to operationalize 'the public' as an actor in crime and violence reduction initiatives will have to be tested. Transition processes in El Salvador, Honduras, Ecuador, Brazil, and Belize have gathered important experience in different shapes and forms of public participation, and they could serve to inform future efforts at public participation in negotiations with CAGs. A more innovative option could be to develop a system of balloting, i.e. to have members of a public negotiating team chosen by lot. Although public participation in public policymaking and decision-making is still not much practised in Latin America, there are some experiences to draw upon. Ecuador, for example, integrated this practice into its 2008 constitution: it states in Article 209 that the members of the Council of Citizen Participation and Control should be elected through sortition. Van Reybrouck (2016) in his book *Against Elections* discusses a range of different initiatives in which sortition was used to create working groups to develop advice or legislation on specific topics. Sortition could therefore be a model to use to choose members of a negotiating team on behalf of the public. Another model would be for neighbourhood assemblies to choose 'delegates', as is done in Porto Alegre, Brazil, as part of the city's participatory budgeting (Smith and Fletcher 2016).

A key question will be how the involvement of 'the public' can foster the legitimacy of a process and provide 'social support' to sustain negotiations with CAGs. This is especially important when a certain level of criminal activity continues, as well as when implementation of an agreement becomes contested (Cockayne 2013). Participation can initiate a wider dialogue on issues of social exclusion and stigmatization, enabling CAG members to reconnect with a society from which they have withdrawn and which they perceive as largely hostile (and vice versa) (Tager and Umaña 2013; Seelke 2016). If parties can work through significant periods of tension it can create a stable foundation for implementation and a realistic view about a possible 'way out' of crime and violence.

Another important point is the need for objective information and careful communication to guide a negotiation process with CAGs. Communication should follow the example of Nelson Mandela in South Africa, who used language that helped to heal the nation rather than further antagonize it. With regard to CAGs, attention to credibility and honour is important, as these are significant attributes which all those involved in a transition process have to internalize to structure their communication and interaction. To limit the impact of disinformation, negotiation parties can opt to create a credible news, information, or social media platform to share reliable information about the process, possibly through the innovative use of blockchain technology (Bock 2012; Seelke 2014). Third parties could play such roles, including news agencies, other non-state actors, and local civil society initiatives (Briscoe *et al.* 2014). It is also interesting to look at the way information is shared with neighbourhoods in the participatory budgeting process in Porto Alegre, where interaction with the public, who participate directly in allocation decisions, is continuous throughout the year using three types of meetings: neighbourhood assemblies, 'thematic' assemblies, and meetings of delegates for citywide coordinating sessions (Smith and Fletcher 2016). This system allows a constant information flow and, depending on the context, could be an interesting model to adapt for negotiations with CAGs, possibly to complement other platforms.

Strengthening justice institutions and accountability mechanisms and fighting impunity are critical elements to avoid people falling back into armed violence and crime (Parenti 2000; Sampaio 2014). These elements need to be addressed from the beginning of negotiated exits from crime and violence: all parties have to accept the overall aim to return to democratic values, but exactly the acceptance of this aim is often difficult in contexts where years of chronic violence have left deeps scars in societies that can be accompanied by harsh opinions about how 'criminals' should be treated and what they deserve. The Proyecto de Opinion Publica de America Latina (LAPOP) survey showed that support for the break with democracy due to criminal violence is the highest precisely in those countries where violence is highest (Adams 2014). Due to high crime rates and very low conviction rates, people have low levels of trust in law enforcement, which has in turn increased support for government initiatives aimed at increasing the role of the military in public security (ICG 2017). The LAPOP data show that

> … those who have been victims of crime or who perceive that crime is increasing in their countries express less support for the political system and the rule of law than other citizens, including less support for the idea that police should always obey the law.
>
> (Cruz 2008: 232)

Creating more political space for resolving criminal armed violence through negotiations could be a good starting point to revise this trend (Canton 2015). Other issues to address are conditions of detention inside prisons, the treatment of detainees, and the high levels of overcrowding (US Department of State 2013; UNHRC 2013; Bergman *et al.* 2015). Many members of CAGs will in the end remain incarcerated, which is something the public demands and most CAG members may be able to accept as an outcome if prison conditions improve (Piché 2016).

Considering that CAGs will probably have no experience in official negotiation procedures, they will need capacity building for participation in negotiation processes. As with NSAGs, such support can be provided by a neutral third party which can also accompany a CAG during setbacks in the process and eventually when implementing an agreement. Religious actors could play a role because they often have the trust and confidence of both state authorities and CAGs (Whitfield 2013; Dudley 2010). An international fourth party as an observer or witness, like the Organization of American States in the case of El Salvador, can also help foster further goodwill and mobilize funds needed for implementation of an agreement (Armengol 2013).

Identifying sponsors and spoilers of negotiations and, where possible, limiting their capacity to harm the process are key tasks during negotiations. Both will either have to be coopted as much as possible or see their impact neutralized. Much like in post-conflict situations, interests of elites have to be catered for (Valters *et al.* 2015; Stedman 1997). Other third parties, including private businesses, political parties, and media outlets, could play an important role in this respect. Religious entities, such as the Catholic Church or the Pentecostal Church, can have influence over actors, sponsors, and spoilers. In fact, for gang members 'religious awakening' is one of few reasons they have to leave a gang (Zaluar 2010).

Within the overall approach to negotiations, trauma and anxiety must be taken into account as part of the impact of chronic violence on a population, including perpetrators. This is for two main reasons. First, on a purely technical level, trauma affects reasoning processes on all sides, and as such can severely impact negotiations with CAGs (Ward 2012, Adams 2014). Second, the aim of entering into negotiations with CAGs is not merely to work things out with one group, but to use the space created by these negotiations to address the underlying drivers of chronic violence and thereby work towards sustainable outcomes. Ignoring this would simply lead to exchanging one group for another. Including suffered trauma and anxiety of a society or community in the larger negotiation frame can help create a 'new normal' in which a spiral of constructive peaceful solutions replaces the pre-existing destructive cycle of continuously escalating violence.

Finally, labelling groups as 'criminal', although legally correct, does not promote dialogue and negotiation processes. National and international legal restrictions on engagement with CAGs should be suspended for the duration of a negotiation process that aims to reduce criminal armed violence. An example is the suspension of arrest warrants for members of FARC–EP while formal negotiations to end Colombia's civil war were under way in Cuba (Kemp and Shaw 2014; Sloan and Cockayne 2014).

Conclusion

Overcoming the moral hurdles and prejudices associated with negotiating with CAGs as a means to reduce violence and crime is a complicated journey in a state-centric world; yet it appears similar to the one taken in the course of the twentieth century when states overcame their objections to negotiate with liberation movements and NSAGs. It remains difficult to free ourselves from the shackles of indignation and fear related to CAGs, even if it could help urban communities move away from the chronic violence that surrounds them. Moving towards an objective assessment of the possibilities to exit criminal violence, at times to the detriment of powerful individual or group interests with a stake in the status quo, takes courage, strength, and political will. These ingredients open new policy space to enable opportunities for negotiations with CAGs that forge lasting exits from violence and more inclusive and safer countries and cities.

In the end, negotiating with CAGs is essentially no different from negotiating with NSAGs, and there is therefore no need for complex new handbooks or guidelines. Much of existing negotiation and mediation practice can be translated for negotiations with CAGs. The risks of engaging CAGs should be taken seriously, as is the case with any engagement of an NSAG; but they should also not be exaggerated as a major deviation from decades of negotiation practice. After all, dealing with criminal rents has been part and parcel of many peace process negotiations in one way or the other, and should therefore not be seen as something new or impossible to do.

If we are serious about building sustainable and safe cities, it is difficult to forgo engaging with CAGs and the enablers that allowed them to grow into the formidable threat to urban safety they have become. As such, this chapter's objective is to develop an argument for integrating negotiations with CAGs into the violence reduction toolbox as part of a larger urban peacebuilding agenda. It suggests four main conclusions.

First, negotiations with CAGs need to be approached in the same way as negotiations with NSAGs. Whether or not to negotiate with CAGs should depend on objective criteria, including those related to ripeness and the issue of stalemate discussed in this chapter.

Second, negotiations cannot solely focus on ending crime or violence; they should be part of a larger urban change agenda. Experiences from different Latin American countries suggest that focusing only on violence reduction risks underestimating the multidimensional nature of CAGs and the power of social exclusion, unemployment, state repression, and stigmatization to embed chronic violence.

Third, civil society in cities can become an important actor in its own right in negotiation processes, compensating for the impact of corruption and impunity and other sources of bias against negotiation. In any case, a discussion about societal change cannot occur without civil society, especially urban populations most affected by armed or criminal violence. Finding a modern way of integrating civil society as an actor into such processes from the beginning could make a real difference in improving safety, and as such provide a space for more sustainable peacebuilding outcomes with more resilient and empowered urban communities that have become part of the solution.

Finally, access to objective information and the use of professional communication strategies are crucial for negotiations with CAGs. Communication platforms that are created for a peace processes could persist to inform, for instance, city councils or neighbourhood communities, and thereby assist in implementing agreements and facilitate dealing with future issues as they arise as part of a transformation process.

References

Abbate, L. and P. Gomez. 2007. *I Complici: Tutti Gli Uomini di Bernardo Provenzano da Corleone al Parlamento.* Rome: Fazi Editore.

Adams, T. M. 2014. *Chronic Violence and Non-Conventional Armed Actors: A Systemic Approach.* Oslo: NOREF.

Algar-Faria, G. 2013. *Understanding the Conditions Necessary for Fruitful Negotiations in Afghanistan.* London: Foreign Policy Centre.

ALRES (Asamblea Legislativa de la Republica de El Salvador). 2003. 'Decreto 158: Ley Anti-Maras', 10 October.

ALRES (Asamblea Legislativa de la Republica de El Salvador). 2010. 'Decreto 458: Ley de Proscripción de Maras, Pandillas Asociaciones y Organizaciones de Naturaleza Criminal', 9 September.

Alvarez, E. and T. P. Prado. 2002. 'Guatemala's Peace Process: Context, Analysis and Evaluation', in C. Barnes (ed.) *Owning the Process: Public Participation in Peacemaking*, ACCORD 13. London: Conciliation Resources.

AP (Associated Press). 2010. 'Mexican Cartels Move Beyond Drugs, Seek Domination', 8 April, available at www.nbcnews.com/id/38565051/ns/world_news-americas/t/mexican-cartels-move-beyond-drugs-seek-domination/#.Wyf3UO8UkqM (accessed 6 July 2018).

Armao, F. 2014. 'Criminal Clusters: State and Organized Crime in a Globalized World', *European Review of Organized Crime* 1(1), 1–44.

Armengol, V. F. 2013. *The Principles of Mediation and the Role of Third Parties in Peace Processes.* Oslo: NOREF.

Ballentine, K. and H. Nitzschke. 2005. *The Political Economy of Civil War and Conflict Transformation.* Berlin: Berghof Research Centre for Constructive Conflict Management.

Bangerter, O. 2010. 'Territorial Gangs and Their Consequences for Humanitarian Players', *International Review of the Red Cross* 92(878), 400–401.

Barton, R. L. 2012. 'The Role of Victims in Plea Bargaining', *The Crime Report*, 5 March, available at http://thecrimereport.org/2012/03/05/2012-03-the-role-of-victims-in-plea-bargaining/ (accessed 13 June 2018).

Bensman, J. 2015. 'Max Weber's Concept of Legitimacy', in R. Jackall and D. Graham (eds) *From Joseph Bensman: Essays on Modern Society.* Knoxville, TN: Newfound Press, 325–367.

Bergman, M., L. E. Amaya, G. Fondevilla, and C. Vilalta. 2015. *Reporte de Cárceles en El Salvador*. San Salvador: Universidad Francisco Gavidia.

Bock, J. G. 2012. *The Technology of Nonviolence: Social Media and Violence Prevention.* Cambridge, MA: MIT Press.

Bodrero, L. 2013. 'Italian Trial Looks at Mafia–State Pact', Organized Crime and Corruption Reporting Project, 14 May, available at www.occrp.org/en/investigations/1955-italian-trial-looks-at-mafia-state-pact (accessed 13 June 2018).

Briscoe, I. 2013. *Non-Conventional Armed Violence and Non-State Actors: Challenges for Mediation and Humanitarian Action.* Oslo: NOREF.

Briscoe, I. 2015. 'Crime and the State: Latin America's Season of Scandal', 26 February. The Hague: Netherlands Institute of International Relations Clingendael.

Briscoe, I., C. Perdomo, and C. U. Burcher (eds). 2014. *Illicit Networks and Politics in Latin America.* Stockholm and The Hague: International IDEA, Netherlands Institute of International Relations Clingendael, and NIMD.

Bustelo, M. G. 2015. 'El Salvador's Gang Truce: A Lost Opportunity?', *Open Security*, 18 May, available at www.opendemocracy.net/opensecurity/mabel-gonz%C3%A1lez-bustelo/el-salvador%E2%80%99s-gang-truce-lost-opportunity (accessed 13 June 2018).

Canton, R. 2015. 'Crime, Punishment and the Moral Emotions: Righteous Minds and Their Attitudes Towards Punishments', *Punishment and Society* 7(1), 54–72.

Cantor, D. J. 2014. 'The New Wave: Forced Displacement Caused by Organized Crime in Central America', *Refugee Survey Quarterly* 33(3), 43–68.

Carpenter, A. C. 2016. 'Gang Conflict and Drug Trafficking: Challenges and Opportunities for Conflict Resolution in Mexico and Central America', in K. Howard and J. H. Peterson (eds) *Linking Political Crime and Violence in Latin America: Myths, Realities and Complexities*. London: Lexington Books, 27–50.

Cockayne, J. 2013. *Strengthening Mediation to Deal with Criminal Agendas*. Geneva: Centre for Humanitarian Dialogue.

Corchado, A. and K. Krause. 2016. 'Deadly Deal: A Drug Kingpin's Plea with the U.S. Triggered Years of Bloodshed Reaching All the Way to Southlake', *Dallas Morning News*, 14 April, available at http://interactives.dallasnews.com/2016/cartels/ (accessed 13 June 2018).

Cruz, J. M. 2008. 'The Impact of Violent Crime on the Political Culture of Latin America: The Special Case of Central America', in M. Seligson (ed.) *Challenges to Democracy in Latin America and the Caribbean: Evidence from the Americas Barometer 2006–2007*. Nashville, TN: Vanderbilt University Press, 219–243.

Cruz, J. M. 2011. 'Criminal Violence and Democratization in Central America: The Survival of the Violent State', *Latin American Politics and Society* 53(4), 1–33.

Cruz, J. M. 2015. 'Police Misconduct and Political Legitimacy in Central America', *Journal of Latin America Studies* 47(2), 251–283.

Cruz, J. M. and A. Duran-Martínez. 2016. 'Hiding Violence to Deal with the State: Criminal Pacts in El Salvador and Medellin', *Journal of Peace Research* 53(2), 197–210.

Cure Violence. 2016. 'Report on the Cure Violence Model Adaptation in San Pedro Sula, Honduras', November, available at http://cureviolence.org/wp-content/uploads/2016/08/Report-on-the-Cure-Violence-Adaptation-in-San-Pedro-Sul.pdf (accessed 13 June 2018).

Daugherty, A. 2015. 'El Salvador Supreme Court Labels Street Gangs as Terrorist Groups', *InSightCrime*, 25 August, available at www.insightcrime.org/news-briefs/el-salvador-supreme-court-labels-street-gangs-as-terrorist-groups (accessed 13 June 2018).

Deanton, J. and O. R. Ferreira. 2015. *Detention Without Charge: The Use of Arraigo for Criminal Investigations in Mexico.* San Diego, CA: University of San Diego, Department of Political Science and International Relations.

Dickie, J. 2013. *Mafia Republic, Cosa Nostra, 'nDrangheta, Camorra: 1946 to the Present*. London: Sceptre.

Dudley, S. 2010. 'How "Mano Dura" Is Strengthening Gangs', *InSightCrime*, 21 November, available at www.insightcrime.org/investigations/how-mano-dura-is-strengthening-gangs (accessed 13 June 2018).

Dudley, S. 2013. 'Barrio 18 Leader "Viejo Lin" on El Salvador Gang Truce', *InSight-Crime*, 7 June, available at www.insightcrime.org/investigations/barrio-18-leader-viejo-lin-on-el-salvador-gang-truce (accessed 13 June 2018).

Eguizabal, C., M. C. Ingram, K. M. Curtis, A. Korthuis, E. L. Olson, and N. Philips (eds). 2014. *Crime and Violence in Central America's Northern Triangle: How US Policy Responses Are Helping, Hurting and Can Be Improved*, Woodrow Wilson Center Reports on the Americas 34. Washington, DC: Woodrow Wilson Center.

Escobar, C. T. 2014. 'El Uso e Impactos del Arraigo en México', available at http://fundar.org.mx/otrosreferentes/documentos/DocArraigoOK.pdf (accessed 6 July 2018).

Farah, D. and C. Meacham. 2015. *Alternative Governance in the Northern Triangle and Implications for US Foreign Policy*. Washington, DC: Centre for Strategic International Studies.

Gagne, D. 2015. 'How 100 Years of Failed Drug Policy Gave Rise to Mexico's Cartels', *InSightCrime*, 15 April, available at www.insightcrime.org/news-analysis/how-100-years-failed-drug-policy-rise-mexico-cartels (accessed 13 June 2018).

Geneva Declaration Secretariat. 2008. *Armed Violence Prevention and Reduction: A Challenge for Achieving the Millennium Development Goals*. Geneva: Geneva Declaration Secretariat.

Gilna, D. 2014. 'Sentence Reductions for "Snitching" Undermine U.S. Justice System', *Prison Legal News*, 15 September, available at www.prisonlegalnews.org/news/2014/sep/15/sentence-reductions-snitching-undermine-us-justice-system/ (accessed 13 June 2018).

Gossman, P. 2016. 'Afghanistan War Crimes Suspect Comes Home: Hekmatyar Deal an Affront to Victims of Grave Abuses', Human Rights Watch EDT Dispatches, 21 September, available at www.hrw.org/news/2016/09/21/afghanistan-war-crimes-suspect-comes-home (accessed 13 June 2018).

Grillo, I. 2012. *El Narco: The Bloody Rise of the Mexican Drug Cartels*. London: Bloomsbury Publishing.

Haas, R. 1988. 'Ripeness and the Settlement of International Disputes', *Survival* 30(3), 245–246.

Hagedorn, J. M. 2005. 'The Global Impact of Gangs', *Journal of Contemporary Criminal Justice* 21(2), 157–163.

Hazen, J. M. and D. Rodgers (eds). 2014. *Global Gangs: Street Violence Across the World*. Minneapolis, MN: University of Minnesota Press.

Hernandez, A. 2012. *Los Señores del Narco*. Mexico: Random House Mondadori.

Hernández-Anzora, M. 2016. 'Las Maras Después de la Tregua: Discusión Sobre su Impacto Política', Perspectivas 5/2016. San Salvador: Friedrich Ebert Stiftung.

Höglund, K. 2008. *Peace Negotiations in the Shadow of Violence*. Leiden: Martinus Nijhoff.

IACHR (Inter-American Commission on Human Rights). 2013. 'Report on the Use of Pre-Trial Detention in the Americas', Organization of American States, available at www.oas.org/en/iachr/pdl/reports/pdfs/Report-PD-2013-en.pdf (accessed 13 June 2018).

IACHR (Inter-American Commission on Human Rights). 2015. 'Violence, Children and Organized Crime', Organization of American States, OEA/Ser.L/V/II., Doc. 40/15, 11

November, available at www.oas.org/en/iachr/reports/pdfs/violencechildren2016.pdf (accessed 6 July 2018).

IACHR (Inter-American Commission on Human Rights). 2017. 'Report on Measures Aimed at Reducing the Use of Pre-Trial Detention in the Americas', Organization of American States, available at www.oas.org/en/iachr/reports/pdfs/PretrialDetention.pdf (accessed 13 June 2018).

ICG (International Crisis Group). 2017. 'El Salvador's Politics of Perpetual Violence', ICG Latin America Report 64. Brussels: ICG.

Idler, A. 2016. 'Por Qué Colombia Ha Votado No y Qué Significa.' *DemocraciaAbierta*, 4 October, available at www.opendemocracy.net/democraciaabierta/annette-idler/por-qu-colombia-ha-votado-no-y-qu-significa (accessed 13 June 2018).

IDMC (International Displacement Monitoring Centre). 2015. *New Humanitarian Frontiers: Addressing Criminal Violence in Mexico and Central America.* Geneva: IDMC.

Kemp, W. and M. Shaw. 2014. *From the Margins to the Mainstream: Toward an Integrated Multilateral Response to Organized Crime.* New York: International Peace Institute.

Kinosian S., A. Albaladejo, and L. Haugaard. 2016. 'El Salvador's Violence: No Easy Way Out', Latin America Working Group, 10 February, available at www.lawg.org/action-center/lawg-blog/69-general/1578-el-salvadors-violence-no-easy-way-out (accessed 13 June 2018).

Koonings, K. and D. Kruijt (eds). 2007. *Fractured Cities: Social Exclusion, Urban Violence and Contested Spaces in Latin America.* London: Zed Books.

Lajous, A. M. 2014. 'The Constitutional Costs of the War on Drug: Ending the Drug Wars', report of the LSE Expert Group on the Economics of Drug Policy. London: London School of Economics and Political Science.

Le Clerq, J. A. and G. R. S. Lara. 2015. *Indice Global de Impunidad.* Puebla: Centre de Estudios Sobre Impunidad y Justicia, Universidad de las Américas.

Lohmuller, M. 2015. 'El Salvador's Gangs & Prevailing Gang Paradigms in a Post-Truce Context', available at www.insightcrime.org/images/PDFs/2015/LohmullerElSalGangs.pdf (accessed 13 June 2018).

Lucchi, E. 2014. *Humanitarian Actors' Struggle for Access: Impartiality and Engagement with Armed Non-State Actors.* Geneva: Professionals in Humanitarian Action and Protection.

Maloione, M. F. T. and C. B. Malone-Rowe. 2014. 'Transnational Organized Crime in Latin America', in J. Albanese and P. Reichel (eds) *Transnational Organized Crime: An Overview from Six Continents.* Los Angeles, CA: Sage, 57–74.

Martínez, C. and R. Valencia 2017. 'Lo que se le Olvida al Gobierno es que al Matarnos Están Matando a la Gente que los Llevó al Poder', *El Faro*, 9 January, available at http://elfaro.net/es/201701/salanegra/19750/'Lo-que-se-le-olvida-al-gobierno-es-que-al-matarnos-están-matando-a-la-gente-que-los-llevó-al-poder'.htm?st-full_text=all&tpl=11 (accessed 13 June 2018).

Martínez d'Aubuisson, J. J. and C. Martínez. 2016. 'Videos Show FMLN Leaders Offering El Salvador Gangs $10 Mn in Micro-Credit', *InSightCrime*, 29 October, available at www.insightcrime.org/news-analysis/videos-show-fmln-leaders-offering-el-salvador-gangs-10-mn-in-micro-credit (accessed 13 June 2018).

Massicard, E. 2010. '"Gangs in Uniform" in Turkey: Politics at the Articulation Between Security Institutions and the Criminal World', in J. L. Briquet and G. Favarel-Garrigues (eds) *Organized Crime and States: The Hidden Face of Politics.* Houndsmill: Palgrave Macmillan, 47–71.

McDermott, J. 2012. 'Money Runs Out for Belize Gang Truce', *InSightCrime*, 24 December, available at www.insightcrime.org/news-briefs/money-runs-out-for-belize-gang-truce?highlight=WyJiZWxpemUiLCJiZWxpemUncyJd (accessed 13 June 2018).

Mijango, R. 2013. *Tregua Entre Pandillas y/o Proceso de Paz en El Salvador*. San Salvador: Red-Imprenta.

Monzón, M. 2017. 'La Criminalidad Compleja', *La Prensa Libre*, 27 June, available at www.prensalibre.com/opinion/opinion/la-criminalidad-compleja (accessed 13 June 2018).

Muggah, R. 2015. 'Fixing Fragile Cities: Solutions for Urban Violence and Poverty', *Foreign Affairs*, 15 January, available at www.foreignaffairs.com/articles/africa/2015-01-15/fixing-fragile-cities (accessed 13 June 2018).

Muggah, R. and I. S. De Carvalho. 2014. 'Changes in the Neighbourhood: Reviewing Citizen Security Cooperation in Latin America', Strategic Paper 7. Rio de Janeiro: Igarapé Institute.

Olson, E. L. (ed.). 2015. *Crime and Violence in Central America's Northern Triangle*. Washington, DC: Woodrow Wilson Center.

Open Society Foundations. 2016. 'Undeniable Atrocities: Confronting Crimes against Humanity in Mexico', 27 June, available at www.refworld.org/docid/5799f7d64.html (accessed 13 June 2018).

Paffenholz, T. 2014. *Broadening Participation in Peace Processes: Dilemmas and Options for Mediators*. Geneva: Centre for Humanitarian Dialogue.

Paley, D. 2014. *Drug War Capitalism*. Edinburgh: AK Press.

Parenti, C. 2000. 'Crime as Social Control', *Social Justice* 27(3), 43–49.

Perdomo, C. and C. U. Burcher. 2016. *Protecting Politics: Deterring the Influence of Organized Crime on Local Democracy*. Stockholm and The Hague: International IDEA and Netherlands Institute of International Relations Clingendael.

Perez-Salazar, B. 2016. 'Hybrid Order: Governance in Rapidly Expanding Urban Areas', in International Centre for Prevention of Crime, *Crime Prevention and Community Safety: Cities and the New Urban Agenda*. Montreal, QC: ICPC, 29–30.

Petrasek, D. 2004a. 'Armed Groups and Peace Processes: Asymmetric Mediation – Take 2', working paper. Geneva: Centre for Humanitarian Dialogue.

Petrasek, D. 2004b. 'Armed Groups and Peace Processes: Pondering and Planning Engagement', working paper. Geneva: Centre for Humanitarian Dialogue.

Phillips, D. 2018. 'Brazilian Army to Take Control of Security in Rio as Violence Rises', *Guardian*, 16 February, available at www.theguardian.com/world/2018/feb/16/brazilian-army-rio-de-janeiro-michel-temer (accessed 13 June 2018).

Piché, G. R. 2016. 'Assessing the Impact of Orthodox Security Sector Reform in El Salvador', CSG Paper 10. Kitchener, ON: Centre for Security Governance.

Planta, K. and V. Dudouet. 2015. *Fit for Negotiation? Options and Risks in the Political Transformation of Non-conventional Armed Groups*. Oslo: NOREF.

Podder, S. 2012. *From Spoilers to Statebuilders: Constructive Approaches to Engagement with Non-State Armed Groups in Fragile States*. Paris: OECD.

Powell, J. 2014. *Talking to Terrorists: How to End Armed Conflicts*. London: Bodley Head.

Reitano, T. and M. Hunter. 2016. *Protecting Politics: Deterring the Influence of Organized Crime on Public Service Delivery*. Stockholm and The Hague: International IDEA and Netherlands Institute of International Relations Clingendael.

Rexton Kan, P. 2015. 'Mexican Cartels as Vicious Firms', *Small Wars Journal*, 15 March, available at http://smallwarsjournal.com/jrnl/art/mexican-cartels-as-vicious-firms (accessed 13 June 2018).

Sampaio, A. 2014. 'Out of Control: Criminal Gangs Fight Back in Rio's Favelas', *Jane's Intelligence Review* 26(12), 44–48.

Santamaria, G. 2014. *Drugs, Gangs, and Vigilantes: How to Tackle the New Breeds of Mexican Armed Violence.* Oslo: NOREF.

Schirch, L. 2009. 'Structuring Public Participation in Official Peace Processes. Afghanistan: Pathways to Peace', Policy Brief 4. Washington, DC: 3D Security Initiative.

Schultz, R., D. Farah, and I. Lochard. 2004. 'Armed Groups: A Tier-One Security Priority', INSS Occasional Paper 57. Colorado Springs, CO: US Air Force Institute for National Security Studies.

Seelke, C. R. 2014. *Gangs in Central America.* Washington, DC: Congressional Research Service.

Seelke, C. R. 2016. *Gangs in Central America.* Washington, DC: Congressional Research Service.

Sloan, B. and J. Cockayne 2014. *Terrorism, Crime and Conflict: Exploiting the Differences Among Transnational Threats?* New York: Centre on Global Counter-Terrorism Cooperation.

Smith, S. and S. Fletcher 2016. 'Participatory Budgeting: Porto Alegre', *Participedia.* 27 May, available at http://participedia.net/en/cases/participatory-budgeting-porto-alegre (accessed 13 June 2018).

Stedman, S. J. 1997. 'Spoiler Problems in Peace Processes', *International Security* 22(2), 43–52.

Susskind, L. 2010. 'Complexity Science and Collaborative Decision Making', *Negotiation Journal* 26(3), 367–370.

Tager, A. G. and I. A. Umaña. 2013. *La Tregua entre Pandillas: Hacia un Proceso de Construcción de Paz Social.* Guatemala City: Interpeace.

The Economist. 2016. 'A Tumultuous Final Chapter', *The Economist,* 10 December, available at www.economist.com/news/americas/21711526-governments-pragmatic-decision-not-call-second-referendum-comes-cost-colombias (accessed 13 June 2018).

Torrealta, M. and G. Mottola 2012. *Proceso allo Stato.* Milan: BUR Rizzoli.

Umaña, I. A., B. Arevalo de Léon, and A.G. Tager. 2013. 'El Salvador, Negotiating with Gangs', ACCORD 25. London: Conciliation Resources.

Ungar, M. 2009. 'Policing Youth in Latin America', in G. Jones and D. Rodgers (eds) *Youth Violence in Latin America.* Houndsmill: Palgrave Macmillan, 203–224.

UNHRC (UN Human Rights Council). 2013. 'Report of the Working Group on Arbitrary Detention Addendum: Mission to El Salvador', UN Doc. A/HRC/22/44/Add. 2, 11 January, available at www.un.org/ga/search/view_doc.asp?symbol=A/HRC/22/44/Add.2 (accessed 13 June 2018).

United States of America v. *José Rafael Abello Silva.* 1990. Reporter's transcript of proceedings, 6 April, available at http://nsarchive.gwu.edu/NSAEBB/NSAEBB2/docs/doc17.pdf (accessed 6 July 2017).

UNODC (UN Office on Drugs and Crime). 2012. *Delinquencia Organizada Transnacional en Centroamérica y el Caribe: Una Evaluación de Amenazas.* Vienna: UNODC.

US Department of Justice. 2012. 'A Review of ATF's Operation Fast and Furious and Related Matters.' Washington, DC: Office of the Inspector General Oversight and Review Division.

US Department of State. 2013. 'El Salvador 2013 Human Rights Report: Country Reports on Human Rights Practices for 2013.' Washington, DC: Bureau of Democracy, Human Rights and Labor.

Valters, C., E. Van Veen, and L. Denney. 2015. 'Security in Post-Conflict Contexts', Dimension Paper 2. London: Oversees Development Institute.

van der Borgh, C. and W. Savenije 2015. 'De-Securitising and Re-Securitising Gang Politics: The Funes Government and Gangs in El Salvador', *Journal of Latin American Studies* 47(1), 149–176.

van Reybrouck, D. 2016. *Against Elections: The Case for Democracy*. London: Bodley Head.

Velez, J. C. 2001. 'Conflicto y Guerra: La Lucha por el Orden en Medellin', *Estudios Politicos* 18, 61–89.

Vivanco, J. M. 2016. 'Peace and Justice in Colombia', Human Rights Watch, 14 October, available at www.hrw.org/news/2016/10/14/peace-and-justice-colombia (accessed 13 June 2018).

Ward, T. W. 2012. *Gangsters without Borders: An Ethnography of a Salvadorian Street Gang*. Oxford: Oxford University Press.

Wennmann, A. 2014, 'Negotiated Exits from Organized Crime? Building Peace in Conflict and Crime-Affected Contexts', *Negotiation Journal* 30(3), 255–273.

Whitfield, T. 2013. *Mediating Criminal Violence: Lessons from the Gang Truce in El Salvador*. Geneva: Centre for Humanitarian Dialogue.

Wolff, M. J. 2015. 'Building Criminal Authority: A Comparative Analysis of Drug Gangs in Rio de Janeiro and Recife', *Latin American Politics and Society*, 57, 21–40.

World Bank. 2011. *World Development Report: Conflict, Security and Development*. Washington, DC: World Bank.

Yawanarajah, N. 2016. 'Ripe, Ready or Strategic? The Timing of Peace Initiatives', Oxford Research Group, available at www.oxfordresearchgroup.org.uk/sites/default/files/Ripe,%20Ready%20or%20Strategic%20October%202016_0.pdf (accessed 13 June 2018).

Zaluar, A. 2000. 'Perverse Integration: Drug Trafficking and Youth in the Favelas of Rio de Janeiro', *Journal of International Affairs* 53(2), 653–671.

Zaluar, A. 2004. *Integração Perversa: Pobreza e Traffico de Drogas*. Rio de Janeiro: Editora FGV.

Zaluar, A. 2010. 'Turf War in Rio de Janeiro: Youth, Drug Traffic and Hyper-Masculinity', *Vibrant – Virtual Brazilian Anthropology* 7(2), 17–23.

Zartman, I. W. 2001. 'The Timing of Peace Initiatives: Hurting Stalemates and Ripe Moments', *Global Review of Ethnopolitics* 1(1), 8–18.

Zartman, I. W. 2005. 'Looking Forward and Looking Backward on Negotiation Theory', in I. W. Zartman and V. Kremenyuk (eds) *Peace Versus Justice*. Oxford: Rowman and Littlefield, 289–301.

5 'Pay up or get hurt'

What extortion says about urban governance, and how it might be curbed

Tuesday Reitano and Mark Shaw

Introduction

In large urban areas that have been afflicted by conflict or where there have been sustained periods of high crime and violence, the mode of governance is often one administered by criminal groups. The starkest indicator of this is the strength of the local extortion economy: businesses and ordinary people effectively being asked to pay for protection, and liable to suffer injury or loss from the groups they are paying if they do not. Such an economy develops in places where organized crime groups and gangs are present and systems of law enforcement, local governance, and community cohesion are weak.

Across the globe, business zones where extortion is prevalent visibly show the impact: empty or closed shops, lack of investment, a feeling of unease among business owners, and high levels of outward migration among entrepreneurs and those threatened. Small businesses, which have few resources to absorb the costs of threats, violence, and damage to property, are particularly affected, as extortion markets are generally carefully calibrated to ensure businesses do not always collapse entirely but struggle to make viable profits. This is particularly the case in developing countries, where the phenomenon has received comparatively little analysis.

This chapter shows that the nature of local extortion differs from place to place. In some cities extortion may be limited to only a few economic sectors, mainly in the nighttime economy where illicit transactions around drugs and sex cannot be guaranteed by the state – with criminal groups effectively fulfilling this task. But in other places extortion has become a pervasive reality governing all aspects of daily life: homes, businesses, investments, and infrastructure are extorted for 'rent'; and people trying to cross borders, board public transport, or even access their schools or workplaces are intimidated to pay the 'toll'.

Why is extortion important to understand?

Extortion – effectively paying a 'tax' to avoid violence – is at its heart a form of criminal governance. In some respects it is the ultimate mafia crime, a mechanism in which criminal control is exerted and legitimate forms of governance are literally squeezed out. Extortion is and has been a central practice of the Italian Mafia, and some of the most important analytical work has focused on the associated extortion or protection economies in Italy (see for example Gambetta 1993). What the Italian and other examples suggest is that once extortion economies are established and functioning effectively, they are very difficult to eradicate – that fact alone representing the extent to which the practice of illegal extortion is able to subvert legitimate forms of governance.

The practice of extortion is widely connected in the literature to the formation of states – effectively the collection of a tax in exchange for the provision of collective security (Tilly 1985; Varese 2001). In that sense, criminal extortion where state institutions are present is a practice in direct competition with the 'taxation-state, since the objective to which [criminal groups] aspire is very much the same: To systematically and coercively extract surplus economic capital from a group of productive people dwelling within a particular area' (von Boemcken 2013: 151). Yet even in this relationship of potential or apparent competition, the reality is often that patterns of extortion exist along a spectrum. In areas heavily dominated by gangs, protection payments may be essential (and even an accepted practice) for local businesses and households to maintain their security. Practices of extortion are also not only confined to static geographies but are often linked to taxing the movement of goods at particular chokepoints or in territories where criminal control is strong (Shaw 2016a). Finally, extortion can be found in specific parts of the economy even in very strong states with effective systems of 'enforcement', most notably in the nighttime economy and local drug markets (see Hobbs *et al.* 2003).

In all these cases, the practice of extortion is aligned or linked in some way to the state, in terms of its weakness or readiness to compromise. Tacit agreements – particularly in major cities and in the nighttime economy – may exist between the authorities and those who practise extortion: payment is tolerated if violence is kept to a minimum. In places where the writ of the state is weaker, forms of extortion through 'taxation' may also occur with state knowledge, and may well involve individuals or substantial groups within the state security apparatus, who themselves take a cut. For extortion to occur, however, what must generally be present is a body of people *with the ability and demonstrable willingness* to inflict violence on those who will not pay (Volkov 2002). Extortion is thus seldom entirely non-violent: it relies on periodic (one might even say symbolic) violence to demonstrate what the system is capable of.

In its purist (and in the literature most theoretical) form, the primary defining feature of extortion is that it is a practice that only serves the extorters – a protection racket existing to enrich itself (Skaperdas 2001: 174) or to secure monopoly control over key markets (Schelling 1984: 161). Nevertheless, and depending on the context, extorters frequently blur the lines between the legitimate and the illegitimate, the state and the non-state: those who practise it seek to balance a degree of stated acceptability ('we provide a service') with the capability (and demonstrated ability) for unpredictable violence. The language and symbolism of 'service' (even of helping the authorities) often emerge in interviews with both victims and extorters. It is for this reason that some industries, most notably private security, nightclub bouncers, and militia-style forces who promise to deliver safety, often occupy a space where the extorting of payments (or additional payments) is cloaked by a gown of legitimate service delivery (see for example Gambetta and Reuter 1995).

This suggests that a neat distinction between 'legitimate' and 'illegitimate' forms of governance in the discussion on extortion is sometimes misleading (Anzola 2016). The cycles in which extortion economies develop or are displaced are understudied. The splintering of extortion groups provides the terrifying possibility for the victim of being caught in a negative perpetual cycle of extortion, being preyed upon by several competing or cooperating operations within a comprehensive and parasitic protection racket (Schelling 1984: 161).

Who may be involved thus depends heavily on a set of often highly local factors, and the history of extortion in each of the three case studies we present suggests the degree to which historical and contextual factors shape the nature of extortion practices and their participants. As suggested, these often have more to do with a lack of trust in the state, or weak state institutions, then with 'bad' actors displacing 'good' ones. Nevertheless, the costs of extortion for urban governance and local economic development are severe.

Most pertinently, extortion removes cash from local businesses that could be reinvested in their growth – although as a prominent extortioner in one city reminded us, the same applies to government tax! It does appear that rates of payment are calibrated by those demanding them, and there is space for negotiation. But where extortion is the most damaging is that these payments may vary: in poorly regulated systems there is competition as to who must be paid, and new businesses are often fearful of moving into areas where extortion is commonly understood to occur.

As a driver of insecurity in this way, the impact of extortion can be far more than just economic, particularly when people are poor and the scope to extract criminal rents is low. Extortion uses violence and the threat of violence to derive financial returns. If it is targeted at local households or very small businesses – such as street hawkers – the costs may be higher than when well-funded companies are targeted. An important argument here to counter the one given to us by the extortioner above is that the 'protection' of small businesses in well-regulated extortion markets is subsidized by taxing wealthier and larger concerns. Unregulated extortion subverts that by making the little person pay directly for his/her own protection – not to mention in a way that is unaccountable (Schelling 1984). Where people struggle to pay and the threats become more bombastic, the price of life can literally go down. Extortion thus introduces and sustains a culture of insecurity that reverberates across all of society, and often targets the most vulnerable.

Extortion remains one of the most difficult forms of crime to combat. This exchange of cash (or other resources) for 'protection' is underresearched, partly because the people caught in a protection economy live constantly in fear: few are willing to speak about how these arrangements function, and even fewer are prepared to report them. That is one reason why so little research has been done on extortion practices. Armed groups and criminal organizations are vigilant in protecting the sources of their funds, and quick to exact retribution from those who betray them. Moreover, the threat of violence often coerces those in state positions to turn a blind eye, or become complicit in the business. Local officials often deny that extortion exists, even though they know it does, arguing (with a straight face) that it is not reported. Thus in many places an equilibrium is established: extortion is ignored as long as it remains hidden and its violence contained (Anzola 2016; Gambetta 1990; Varese 2014). For this reason publicity is a key response to extortion – and this is also why the criminal groups involved work actively to prevent it.

Extortion is often subsumed within the broader phenomenon of violent gang cultures, conflict, or violent extremism, and is therefore often considered as a secondary concern of policymakers whose primary interest is in reducing levels of violence. But as these payments can constitute crucial resource flows which bolster criminal governance in urban contexts, and given the intrinsic link to violence and social control, protection economies should receive much more attention from those engaged in urban governance.

Extortion is a causal driver of violence and a defining feature of urban criminal governance in the Central American cities that now globally experience the highest levels of crime and homicide. In this region, perhaps more than any other place on Earth, an all-pervasive and violent extortion economy has entrenched itself, and it is here that we turn for our first case study.

Money or power?

In Central America, and in particular in the three Northern Triangle countries of El Salvador, Honduras, and Guatemala, extortion presents a widespread and pernicious challenge. The consequences for the economy, security, and development have been manifold. In 2017 El Salvador's homicide rate was 60 per 100,000 population – one of the highest in the world – followed by Honduras with a rate of 42.8 and Guatemala with 26 per 100,000 population (Igarapé 2017). Many of these deaths are attributed to street gangs and their practice of violent extortion. For communities that are already poor, extortion has cheapened the cost of human life.

Extortion has also constrained Central America's economies and the region's global competitiveness. In the latest Global Competitiveness Index issued by the World Economic Forum (WEF 2017), the countries in Central America ranked at the bottom due to organized crime's impact on business. A number of studies have estimated the high costs to businesses and the economy to operate in situations of continuous crime and violence. A systematic effort undertaken by the Inter-American Development Bank across Latin America and the Caribbean suggested that in 2014 for the Central American countries the private costs of crime were approximately 2 per cent of gross domestic product (GDP) (Jaitman and Torre 2017: 23).

Extortion profits have become the lifeblood of organized crime in the region: extortion is one of the main funding sources for organized crime groups and street gangs, and in Honduras it is the foremost revenue stream for Barrio 18 and the Mara Salvatrucha, the two largest of the region's gangs. They use these revenues to increase further their economic and social control over communities and municipalities, in an ever-tightening Gordian knot (ICG 2017a).

Beyond mere economics, the social cost is also huge and has intergenerational impact (Adams 2011: 21). While women are a minority in Central American gangs, they are nonetheless central to gang operations. An estimated 20–40 per cent of *maras* are women, and they are often assigned responsibility for the collection of extortion payments. Life in gang-controlled communities is harsh for women, and characterized by high levels of gender-based violence and rape. Gangs in Central America can also be involved in human trafficking to service demand in prostitution (ICG 2017a). Threats against female family members are used to coerce and threaten local police, authorities, rival gangs, and subordinates. A recent study found that threats and murder were the two main factors driving forced displacement in the Northern Triangle, resulting in more than 700,000 displaced persons across the region, of whom women and children make up the largest proportion (Beltrán 2017).

In El Salvador the problem is most acute: extortion as a practice is an integral part of the local identity for gang members, where the status given by their membership is more important than the profits (or lack thereof) to be earned in the illicit economy. Gangs' presence at the municipal level in El Salvador is pervasive; in 2016 the Central Bank estimated that 70 per cent of businesses in El Salvador are subject to systematic extortion, and they paid around US$756 million in extortion payments in that year – the equivalent of 3 per cent of the country's GDP. Everyone is extorted, from international businesses down to the local street vendors. For the most part, however, this is a micro economy: payments demanded rarely exceed US$20 at a time, and the majority may be less than US$5. But missing two of these tiny payments is enough to warrant losing your life (O'Toole 2017).

Significantly, while extortion is the biggest source of revenue for the country's two rival gangs, Barrio 18 and the Mara Salvatrucha (also known as MS-13), the lowest of the rank and file make little or nothing. Like drug dealing, extortion earns few resources for street-based practitioners – money percolates to the top. An investigation by *El Faro* and the *New York Times* found MS-13's estimated earnings from extortion, US$31.2 million a year, would provide a per capita income for each gang member of less than US$65 a month – half the minimum wage of a farmworker. Because the sums being extorted in El Salvador are so small, raids and sweeps by police and justice officials give minimal return on investment. Indeed, between 2012 and 2015 the total amount confiscated by the police in anti-extortion raids was less than US$35,000 (Martínez *et al.* 2016).

In El Salvador, therefore, it is not money that drives the extortion economy, but a desire for control – the transaction around extortion is effectively a symbol of who governs. The gangs use extortion as a sociopolitical show of force, demanding the closure of businesses, including the notorious four-day bus strike in July 2015 that crippled the national economy but netted no revenue for the groups involved. Drivers were told to stay home or pay with their lives. On those four days alone, nine bus drivers were killed for daring to flout the shutdown (Izadi 2015).

These dynamics mean that criminal justice initiatives struggle to gain traction. With sums so small, and the threats to life so potent, few victims bother to report. And as the threats are just as effectively applied to police officers and members of the judiciary who attempt to prosecute, few cases are brought to any kind of meaningful resolution. In data provided by the Salvadoran National Police, between 2013 and 2015 they received only 7,506 reports of extortion leading to just 424 convictions. The extortion's intimacy is inescapable, leaving many victims little choice but to flee, often to the United States, or be killed.

These Central American experiences challenge the understanding that extortion is primarily a money-making enterprise. They show that payments for protection often have a symbolic dimension as part of a broader governance strategy, or drives to control people or markets. In this context, succumbing to demands for extortion payment means agreeing to criminal governance. And given the local nature of gang control and power, and the weakness or absence of state institutions, those targeted for extortion have in most cases no option to desist. These multiple transactions across society in turn constitute a powerful form of local political control: society can be shut down on the orders of those who control the transactions and manage the violent actors associated with their enforcement.

From political donations to protection money

The overlap between politics and criminal governance (through extorting funds) is a critical feature of another powerful example – that of Karachi. Here extortion became the most pervasive of all crimes. No one was spared: bankers, doctors, businessmen, but also teachers, hospital workers, construction workers, and religious figures (Hussain 2016). The city was often described as 'ungovernable'. Much has now changed, but the evolution of the protection economy (and how it has been reduced) says much about the links between politics and crime.

Initially in Karachi collections were made in the name of *zakat* (the donations made to the poor that are one of the five pillars of Islam) or as a political donation. But in the 1990s one of the political parties that was short of funds, the Muttahadi Qaumi Movement (MQM), began to push the concept more aggressively, demanding support in a system that came to be known as *bhatta* or 'protection' (Yusuf 2013). According to estimates at the time, by the end of the 1990s MQM was raising some 765 million Pakistani rupees a year (approximately US$17.4 million at the 1998 exchange rate), and the money was being used for acquisition and storage of weapons, weapons training, funerals of workers who died for the party and the welfare of their bereaved families, medical treatment of injured workers, public relations, and legal defence and buying off witnesses in legal cases against the party (Khan 2010). Due to its success the practice was quickly emulated by other political parties, including smaller political and religious groups. To make the 'requests' for donations a little more compelling, the parties aligned with local crime groups to do the collections for them (where politics ended and crime began was, as elsewhere, often hard to parse), creating a violent synergy between politics and crime in the city (Hussain 2016).

The police blame the sharp increase in extortion on a criminal gang known as the People's Aman Committee (PAC), based in the district of Lyari, one of the most violent and impoverished parts of the city. But the PAC aggressively expanded the practice into new parts of the city, sparking competition with local control groups there. The gang has links to the Pakistan People's Party, which ruled Pakistan almost continuously from its inception in the 1960s until its defeat in the 2013 general elections. The PAC and the MQM were competing for the city's governance, and the means of battle were the protection economy and the politicization of state institutions they could control through violence (ICG 2017b). 'If some political party says they are not involved in the extortion racket, they are lying,' Majyad Aziz Balagamwala, a former president of the Karachi Chamber of Commerce and Industry, was quoted as saying (Hassan 2013).

As the practice and the competition escalated, so did the level of violence. According to a Pakistani journalist reporting in 2011, 'Political activists-criminals knock at your door demanding donations on the pretext of buying party flags to organise celebrations for a sacred event such as Eid Milad-un-Nabi. If you refuse, there are threats, public beatings, or even a bullet' (*Newsline Magazine* 2011). Gang takeovers, score settling, and vendettas caused murder rates to spike – in August 2013, at the height of the violence, there was an average of nine deaths a day in Karachi (ICG 2014).

The relationship between political parties, criminal groups, and violence became symbiotic. Some crime groups morphed into armed wings of the political parties, while others remained quasi-political groups that engaged in land-grabbing and other mafia-style behaviour (Hussain 2016). But over time this has become genuine criminal governance, which is increasingly leeching into the legitimate economy and society. Service provision has been captured by the crime gangs: the 'water mafia' siphons the water supply and sells it to the population for 30 times the government rate. And as the poor settlements with few alternatives often pay the highest rates, sometimes up to a third of their income goes towards water. In 2016 it was estimated that criminal groups were stealing the equivalent of US$540 million worth of water a year, often in collusion with local police (Kunbhar 2016). Similarly, the 'transport mafia' provide minibus and combi services across the city, while at the same time sabotaging normal services and using their political connections to block efforts to build a proper public transportation system for the metropolis. Criminal governance is so widespread in Karachi that ordinary citizens have no choice but to do business with criminals, whether it is to secure drinking water, take a bus, buy a house, or start a business (Shackle 2015). In 2017 the black economy, comprising land, water, and transport mafias, illegal gambling, extortion, human trafficking, and kidnapping, were estimated to be worth US$2.9 billion (ICG 2017b).

Such extortion remains a feature of life in Karachi. Yet what is important to note is that more recently there have been some remarkable success stories in reducing the extent of the protection economy. While we briefly explore this later, it is enough to highlight here that even the most sophisticated and widespread systems of extortion may be curbed – but that a set of conditions need to be in place to make this a possibility. This suggests that the timing of interventions may be crucial, and understanding the evolution of extortion markets constitutes a critical opening for policy analysis.

How extortion evolves

Cape Town is an African city with a thriving extortion economy that has received much attention in research and also more recently in the media due to broader public scrutiny of political corruption (see Pauw 2017, for example). In this city the protection market is particularly complex, drawing on both historical antecedents during the era of apartheid that evolved into a set of established and violent gangs on the city's periphery, and recent trends that include the growth of a private security industry (Shaw and Haysom 2016). From a bird's-eye perspective, Cape Town has three interlinked extortion economies. The first, the mini-bus taxi industry, is not examined here but has a dedicated literature (see Dugard 2001). The main point coming out of this literature is that poorly regulated urban transport systems (as in the case of Central America) are particularly vulnerable to systems of extortion and violence.

The second extortion economy is linked to the evolution of a powerful set of criminal gangs in many marginalized parts of the city. At its most extreme – and especially in neighbourhoods where the police struggle to retain control – the extortion economy resembles the poverty economy described in El Salvador. In Mannenberg just outside Cape Town, for example, an area with a history of significant (although underreported) gang wars, almost all local businesses, both big and small, pay tribute to the gangs. As elsewhere, such payments developed out of the 'tax' on drug dealers, which spread from this illegal economy to the legal one (Petrus and Kinnes 2018). An important point, to which we will return, is that illicit economic regulation (in this case of street-level drug deals) has the ability to mutate, in the absence of effective state responses, to the taxing of all businesses.

Much as the cases of Central America and Karachi demonstrated, the phenomenon of extortion is not new in Cape Town. It occurred in gang-afflicted areas before the forced removals that transformed the city under apartheid. Such payments initially had an element of community solidarity, given the mistrust of the police and the state authorities more widely. That built legitimacy for criminal gangs, which were often themselves important instruments of community engagement and solidarity in a system that had alienated individuals and broken families and other social forms of ordering (Dixon and Johns 2001). These antecedents provided a nascent system of criminal organization that mutated for a variety of reasons, and was often linked to the growth of political parties.

The most significant factor was the sudden and largely uncontrolled flood of cheap drugs into the city from the early 1990s, when South Africa opened up to the world. The sale of cheap drugs was grafted on to an established (but weak) system of protection payment. Each fed off the other as gangs struggled to control the drug trade. The result over the past few years has been violent control over drug market access in the most marginalized areas of the city. Extortion and protection fees align closely with gang control of geographic turf (Shaw 2016b).

A third strand of the extortion economy, and the one that has been most covered in the news of late, is the evolution of the extortion economy in the heart of Cape Town. This probably began in the late 1980s or early 1990s, as a series of private security companies moved into the business of providing protection. How this occurred is the subject of some debate, but what seems clear is that the burgeoning private security industry (South Africa has the largest private security sector in the world) included criminal elements (Goga *et al.* 2014). These coalesced around the protection of the city's nightclubs, taking payments not only for protection but also for controlling the sale of drugs behind the doors. Like its township equivalent, the extortion business was closely linked to the drug trade (Shaw 2016b).

What makes Cape Town an interesting case study is that the second and third strands have evolved over time to intersect with each other. Powerful private security operators have links with violent gangs – providing both legitimate and illegitimate means of enforcement. Until very recently, much of this was not in the news – and indeed there has been evidence of political and police protection of the emerging system (Shaw 2017). The system was largely accepted as established and affecting a wider array of businesses than just nightclubs.

In Cape Town, as elsewhere, a tacit agreement seemed to be in place: no violence or disruption (particularly given the large number of foreign tourists in the city), and things could be left as they were. Violence between those involved in extortion (combined with the efforts of some courageous journalists who began digging to get the story and expose political and police connections) has changed the calculus. Whether the extortion economy will be disrupted for good remains to be seen – it is certainly more exposed than previously. To illustrate just how embedded and fearless the system has become, however, key members involved have continued to collect protection fees while out on bail, thereby effectively thumbing their noses at the state (Thamm 2018).

This is only a brief sketch of a complex extortion economy in Cape Town, but it illustrates several trends. Most importantly, it shows a logic of how extortion markets consolidate over time: different actors and interests converge in ways that reinforce the extortion economy, and these arrangements last as long as there is enough turf and profit to share. When the market for extortion contracts, however, a process of market disruption and recomposition starts whereby certain players are squeezed out. It is these moments of disruption and transition that can provide the most powerful openings for external intervention. Municipal leaders need to be aware of the power balance, recognize the signs of extortion markets in transition, and identify and deploy strategies that will delink local extortion players from political or institutional protection.

The extortion life cycle

If municipalities are going to become sensitive to and prepared for the transitional periods where disruption is possible, there needs to be a nuanced understanding of the evolution of extortion markets. The cases discussed in this chapter illustrate that it is possible to recognize and investigate both the set of preconditions enabling extortion practices and the life cycle over which such practices evolve, intensify, and entrench themselves over time.

Identifying and monitoring this evolution is important, not only because extortion is a crime but also because it is an incremental practice that ratchets up in scope and scale and is often overlooked until it is too late (Anzola *et al.* 2016). In a sense it follows the pattern of the proverbial boiling frog: if a frog is put suddenly into boiling water, it will jump out, but if the frog is put into tepid water which is then brought to a boil slowly, the frog will not perceive the danger and will be cooked to death. If interventions can be made when extortion markets are nascent to prevent the market consolidating in the first place, this is preferable.

Preconditions for extortion: financial need and reservoirs of violence

Two preconditions are typically in place. The first is financial need: where there are limited resources to be earned legitimately, or in situations where they are unequally distributed, extortion becomes a means for some groups to generate rents. Ethnic minorities and stigmatized populations are frequent targets for extortion because of their vulnerability relative to the rest of the population, or the lack of protection and services provided to them by the state (Anzola *et al.* 2016).

Second, extortion requires the presence of reservoirs of violence – groups of actors experienced with the use of violence, and often in contexts where their primary function is no longer required or underutilized. Depending on the context, these can be former combatants, organized crime groups, or military or police personnel. Whether they are the initial proponents of extortion, as with Central America, or are coopted to provide strong-arm services for others, as in Karachi, cannot be determined.

Initial extortion successes

In each case, the extraction of rents that becomes extortion often begins with legitimate or quasi-legitimate appeals for resources to support a specific cause. In El Salvador this took the form of a war tax, in Karachi it was a political tax, and in Cape Town it was rents levied on the expanding drug economy. As the systems evolved, citizens and local businesses become accustomed to making financial contributions; these are initially levied on an extraordinary basis, but then normalize into a standing tax over time.

Expansion and competition

As the case study of Karachi showed, practices of extortion, once proven successful, catch on and spread. This phenomenon is well illustrated by a recently highlighted example in Mexico, where the extortion of avocado farmers has become a chronic and unsustainable problem. According to the Mexican newspaper *Reforma*, in the early 1990s an organized crime group hacked into the computers of the national agricultural ministry (Secretaría de Agricultura, Ganadería, Desarollo Rural, Pesca y Alimentación) and gained access to the records of all avocado growers, including their farm sizes and turnover (Puebla On Line 2017). The Jalisco Neuva Generacion cartel used this information to identify targets and calibrate appropriately the sums they could extort from one of Mexico's most lucrative industries (Ballesteros 2017). A report of the Attorney-General noted that 'The avocado producers were imposed a quota and those that resisted, or the families of those, were kidnapped or murdered as pressure until they ceded' (Fox News 2017). As Jalisco Neuva Generacion proved the success of the strategy, other criminal groups, including the Knights Templar, began to copy the practice, increasing the tax on the avocado industry, which was estimated to be worth US$2.2 billion in 2016 (RT 2017). This criminal tax on Mexico's 'green gold' resulted in more than doubling the market price of an avocado in US markets, where 85 per cent of the annual crop is sold (Ballesteros 2017).

Identity

To practise extortion systematically, the group extorting requires a clear identity so as to apply continued pressure that extends beyond the individual agent. It also serves to band the agents into a network that ensures their loyalty, when they know their efforts could not be successful alone and they themselves could be brought violently into line. They therefore pay their own dues upwards, guaranteeing the longevity of the group.

In the Americas extortion has become intrinsically linked to the gang culture and territorial control, and in Italy the Mafia families use their notoriety and clannishness to inspire fear (Székely *et al.* 2016). A case study of extortion groups in Germany showed that those with the greatest potency used membership of outlaw motorcycle gangs as a defining feature (Anzola *et al.* 2016). In Karachi the association was to political parties and the influence they represented. In Cape Town it was registered security companies with a known brand – albeit with gang connections, as discussed above.

The identity aspect of the extortion economy is what transforms it into a socioeconomic phenomenon and a form of control, rather than being exclusively a financial concern. In some contexts the groups levying an extortion tax will also provide social benefits to their constituent groups; this further enhances the normalization process, as those taxed will see benefits from their payments (Varese 2014). It may also be that the extortionists compare favourably with state institutions that can be similarly, or even more markedly, predatory. In Rio de Janeiro, for example, the rule of the drug lords who extorted the *favelas* and coopted the police units could be less violent and more predictable than the local police, who also widely practised extortion and racketeering in the community (Glenny 2015).

Normalization

Criminal groups use extortion racketeering as a means to control territory, power, and influence by infiltrating the legitimate economy and restricting the activities of rival criminal actors. A well-calibrated extortion strategy exacts a toll, but not too large a toll, on the community – as Frederico Varese (2013: 2) put it in his seminal text, 'mobsters do not overcharge their "victims"'. Businesses can tolerate and absorb the tax; they get used to paying and make payments regularly and on time, thus coming to know their extorter. Studies of extortion in Italy found that groups involved create 'parasitic and symbiotic relationships with their victims' (Transcrime 2009), and that the continuous use of extortion over time makes victims accustomed to it, and progressively collusive with organized crime (Sciandra 2017). In this context levels of violence often go down, as the interactions with organized crime and violent actors become normalized.

Business are increasingly acting unilaterally to address extortion and reduce the violence associated with extortion rackets by seeking direct accommodations with criminal groups, awarding them contracts, and employing gang members (ICG 2017a). Firms in Guatemala City, when the state was unable to intervene effectively to prevent violence and extortion, ended up buying into the protection/extortion system, establishing regular payment systems to gangs to protect themselves from more predatory rivals (ibid.). During the gang truces in El Salvador, some local businesses and communities reached agreements with gangs to offer employment – sometimes legitimate work, sometimes providing 'private protection' – in exchange for lower extortion rates. But these accommodations, where they can be reached, are highly localized (unable to extend beyond the immediate territory of control by that gang) and do not address the long-term damage that criminalized governance will do to the community.

Disruption and reconstitution

The period when competition between groups intensifies and violent conflict breaks out between groups may constitute a critical opening for change, as illustrated by the case of Cape Town. The degree to which that violence is exposed is likely to determine the extent of state responses. In Cape Town, which is heavily reliant on the tourism economy, the response was extreme. Where such conflict remains hidden, perhaps because it is couched within other forms of violence such as conflict or gang behaviour, it is more likely to lead to little or no state responses, or at least not responses that are directly targeted at the extortion economy itself.

Disruption of extortion markets may itself be cyclical. If the same conditions – weak state response, political protection, and the availability of market participants (the extorters themselves) – are in place, it is likely that the market reconstitutes itself, albeit with new players. In fact, absent new conditions, it is almost a guarantee that extortion markets, once established, will reconstitute themselves, only under new leadership. Forging different sets of influences when there are new sets of actors in play can be one opportunity to counteract the extortion economy. It was notable, for example, that the El Salvador gang truces were negotiated in a period when newly emerging criminal groups were displacing the previous 'old guard'. The gang truce was negotiated with the aging cadre of leadership, as a way for them to continue to demonstrate their influence and undermine that of emerging challengers (van der Borgh and Savenije 2015).

Institutional infiltration

The final stage of the extortion evolution is its thorough institutional infiltration. Groups that have successfully wielded extortion as a means of economic and social control can often transition that into the legitimate sphere, buying influence and office, and taking over whole industry sectors and pushing out competitors.

The Japanese Yakuza is perhaps the most advanced and established of groups practising extortion. The Yakuza is well entrenched in social institutions, with historical and resilient links to the Japanese government. Extortion is one of the multiple ways that the group practises social control, and it has become one of the accepted ways to control the country's nighttime and illicit economies, preventing the disruption of businesses and providing conflict resolution. It has also inserted itself into vulnerable points of the legitimate economy, including procedures such as labour negotiations, credit arrangements, dispute resolution, and bankruptcy proceedings, where less ethical businesses may want to tilt decisions in their favour (Anzola *et al.* 2016).

Counter responses becoming new cycles

It is imperative that states respond to extortion, because a failure to do so can result in communities taking matters into their own hands: a privatization of the response. Across the spectrum of examples, communities overweighed by the curse of extortion have created vigilante armies or hire gangs to protect them, often thereby escalating the violence.

Where vigilantism solves one problem, it often creates another. For example, in the Mexican state of Michoacán the vigilante groups created as early as 2010 to break the cycle of drug violence have turned into the next generation of the protection–extortion cycle. In 2016 Mexico's national commission on human rights recommended that the government investigate the integrity of the *autodefensas*, their increasing infiltration by organized crime, and their resistance to demobilization. Without a clear or unifying control structure, the proliferating vigilante groups have become highly politicized, with the groups increasingly falling into conflict among themselves, resulting in sharp increases in local homicide rates (Caballero 2016).

Responses to extortion

Fighting extortion once it is an integral part of the economy and society becomes very difficult. Traditionally, responses to extortion have involved law and order or militarized responses. Yet such approaches have been largely blind with respect to the governance functions of extortion and the associated violence. Responses to extortion have often been categorized as a secondary issue to reducing violence, cracking down on crime, and breaking the gang culture – thus failing to consider the historical links between extortion, gangs, conflict, and political parties.

In the Americas the strategy involved sending the military and police on to the streets to outgun the gangs and fill the country's jails with even the lowest ranking of alleged gang members (Kinosian and Albaladejo 2016). Known as the *mano dura*, these hard-line law enforcement and security strategies resulted in burgeoning prison populations in the region and an increasing emphasis on extortion as a financing strategy for the gangs. The attempt by the El Salvador government to negotiate a gang truce in 2012 resulted in a reduction in the levels of violence over two years, but heightened levels of crime and extortion and gave the gangs a degree of legitimacy they had struggled to achieve before (van der Borgh and Savenije 2015: 168).

The countries in the Northern Triangle are scaling up their efforts to counter extortion. In Guatemala anti-extortion efforts are integrated into the work of the Comision Internacional Contra la Impunidad en Guatemala and the Justice Education Society. In April 2015 they created a specialized office to combat extortion, with separate units dedicated to the strategies and *modi operandi* of different gangs. They have created a permanent hotline to report extortion, and a phone-based app that individuals can use to report and log calls for investigation, and which provides support to victims. The data collected were followed up with three big raids in 2016 that led to more than 200 arrests (ICG 2017a). Municipal efforts to create a lasting and sustainable bulwark to this corrosive practice remain absent, however.

Despite the social rooting of the phenomenon, the responses by states tend to be led exclusively through criminal justice systems, but this approach has been beset by challenges. First, as a former director general of the national police in El Salvador observed at a regional meeting promoting whole-of-community approaches, criminal extortion is structured in such a way that it is rarely an individual activity, but one that is perpetrated and sustained by groups. Thus even when a prosecution of a single criminal or a handful of perpetrators is successful, the group as a whole is able to threaten rapid retribution and individual agents are quickly replaced.[1]

Honduras is considerably stepping up its efforts to counter extortion. Its National Anti-Extortion Force (FNA) was created in 2014, and it has arrested more than 400 people. The government has also joined El Salvador and Guatemala in a tri-national taskforce to counter organized crime. However, their strategy follows a similar criminal-justice-led model that does not integrate in a systematic way with broader efforts to counter insecurity, corruption, and impunity and to protect human rights. These same challenges undermine the response: the FNA has reported that numerous state officials are active participants in extortion rackets. Even where the political will of local government and police is high, death threats against children, wives, and loved ones are a hard challenge for individual officers to resist.[2]

Unlike the Latin America example, where efforts to control the extortion economy came a poor second to the response to gang violence, the government of Pakistan addressed this first as an economic challenge and then as a political one. Karachi was then, and is now, the economic engine of the country. It has long been the nation's financial and industrial hub, and in 2014 was estimated to generate around at least 35 per cent and as much as 50 per cent of the country's gross tax revenue and up to 70 per cent of its GDP. The city is the global trade transport hub, handling an estimated 95 per cent of the country's exports, as well as being the hub for financial services and Pakistan's media capital (Tariq 2015). But in 2013 analysts feared that the extortion economy would smother Karachi's economic growth. Estimates suggested that the extortion threat, coupled with the compromised infrastructure, was causing a 60 per cent decline in domestic investment, as well as capital flight. For the sake of the nation, the situation in Karachi had to be brought under control.

The issue of extortion in Pakistan was literally dealt with as a counterterrorism challenge: the Sindh Rangers paramilitary force was deployed, and the national Counter Terrorism Force was put in the lead. The Rangers were given a fourfold mandate to counter terrorism, assassinations, kidnapping for ransom, and extortion (*Times of Islamabad* 2017). With no civilian oversight bodies or restriction on its duration, the emergency military intervention in what should usually be a civilian policing matter has been extended several times, and is now de facto open-ended and showing evidence of mission creep (ICG 2014).

The militarized enforcement effort has been effective up to a point. The city-wide murder rate has been significantly brought down by as much as 90 per cent from its 2014 high, although critics observe that this number is rather sharply offset by the estimated more than 1,000 people shot dead by security forces under the expanded emergency powers granted to the Rangers to intervene. The Ministry of the Interior reports cases of kidnapping and extortion to have declined by 64 per cent and 73 per cent respectively over the four years between 2013 and 2017, while 533 cases of extortion were reported in the city in 2013 and only 146 in 2016 (Tanoli 2017).

Yet, as we have seen in other jurisdictions, extortion is a crime that is notoriously underreported: the Citizen–Police Liaison Committee (CPLC), a Pakistani non-governmental organization (NGO) committed to improving civilian engagement and oversight in security matters, estimated that at best 30 per cent of cases in Pakistan are reported (Zia 2011). Criminal groups have diversified their activities into different forms of street crime. The most recent CPLC report indicated that bag snatchings, property thefts, and other forms of crime are sharply increasing, continuing the fear with which the citizens of Karachi live (*Business Recorder* 2017). And without alternatives prospects to deliver water or other services to the populace, the reign of the water and transport mafias continues largely unchecked.

In Cape Town, too, the embedded culture of gangs on the margins of the city is key to driving the extortion economy, for both poor and wealthier areas of the city. Unless the culture of gangs can be eroded, undercutting the extortion system will be challenging. Nevertheless, disruption of at least one sector of the extortion system offers hope for a wider set of changes, not least within the private security industry and its links to the criminal underworld. In Cape Town extortion is not a politically neutral zone. As this chapter was finalized, South Africa was in a period of intense political transition following the end of the regime of President Jacob Zuma. This may offer opportunities for breaking patterns of extortion, as key players in these markets have sought and obtained political protection to operate; breaking these national-level links will have local implications for undermining crime governance (Cameron 2009).

New strategies to counter extortion

New approaches are needed to break down and counter extortion economies. Such approaches are an imperative for early and preventive action in countries where awareness-raising and prevention activities could prove catalytic in building community frameworks to counter extortion before it takes hold. In building bulwarks to prevent extortion, there are many lessons and cautionary tales to be learned.

Recognizing extortion as a socioeconomic and political phenomenon, rather than just a security or justice challenge, brings a new set of actors to the table. There is a strong argument for shifting the locus of attention from the national level to the municipal level. Extortion is a strategy of local criminal governance and control, and any sustained response will require strong mobilization and solidarity from key businesses and community figures. It will also require rebuilding trust in the institutions of the state and law enforcement. Building a coalition that spans the private sector, civil society, government, and law enforcement has greater potential as a response framework (Hassan 2013).

Criminal justice responses require a crime first to be reported to allow law enforcement action to be taken, but studies of extortion note that it is a criminal phenomenon that is seriously underreported for a number of reasons. A 1990s' study of extortion by criminal groups in China suggested that it was the most underreported crime, with only a 20 per cent reporting rate (Chin 1996: 96). A national crime survey for Mexico in 2014 found that extortion was the second most common crime citizens experienced, but went 90 per cent unreported (INEGI 2014). Informal conversations by the authors with police officers across the countries mentioned in this chapter suggest that even they deem these low percentages often to be a generous estimate. Fear of retribution, high levels of corruption, the high costs of being involved in a legal suit, and the lack of results have often meant that confidence on the part of the business community in the government and police to address the challenge is low, and they see little return on their investment of increased time, cost, and risk.

Response efforts thus tend to focus first on trying to increase the levels of reporting. Nascent evidence suggests that shifting from individual to collective action and reporting can help to mitigate some of individual barriers and risks of reporting. The most successful campaigns to counter extortion in southern Italy have relied on corporate and consumer unity, and efforts to change the social norms around extortion. For example, an NGO known as Addiopizzo (loosely translating as 'farewell extortion') is fighting extortion by spurring collective action. The group attempts not only to combat the gangs practising extortion, but also to break down the complicity and normalization process that comes as extortion evolves. The group monitors and certifies individual shops as free of extortion, and raises awareness and encourages shoppers only to buy from those shops that have certification. At its core the strategy is trying to change incentive structures, and to build a community of citizens who are ideologically opposed to the Mafia and extortion culture (Székely *et al.* 2016), thereby creating a united front against extortion.

An unexpected success story that similarly demonstrates the benefit of collective action also comes from the traditional heartland of Mafia territory, the city of Palermo in Sicily. The Cosa Nostra branch of the Mafia had all but killed the historic centre of the city. Its unrelenting *pizzo* tax drives businesses away, and local families under pressure progressively abandon the centre, moving to more remote towns where they can live away from the Mafia's power and in greater safety. But in an unexpected quirk of globalization, the city has become a hub for immigration; for refugees and migrants first from Bangladesh and subsequently from sub-Saharan Africa. They have proven remarkably and stoically resistant to the Mafia's threats, even when a migrant was shot at point-blank range by the Mafia for his defiance (Einashe 2016). Whether it is because they cannot afford to or because their tolerance for risk and fear is higher given the often-insecure countries from which they come, the migrants have largely refused to acquiesce to the Cosa Nostra's generations of tyranny (Merelli 2016).

According to Addiopizzio, what made the difference was that the migrants came together as a group to present their case to the police and seek support. Sicilian victims of extortion rarely reported cases of extortion, and if they did so they went individually and quietly. In contrast, the Bangladeshis organized a public street protest, where they marched in solidarity together and then presented a group case to the courts. The migrants' resistance was backed up by the remaining Palermo locals, as well as the local authorities, in a remarkable display of multicultural unity (Einashe 2016).

Another alternative, often suggested in response to persistent criminal markets and the nighttime economy, is to integrate, legalize, and regulate the criminal market itself, thereby eliminating the fluid grey space that extortion groups move into. In response to the Karachi utility mafias, Noman Ahmed (2008), an author and urban planner, suggested that the Sindh government should devise a way to take the water-tanker service into its fold and make it a commercial undertaking that is officially regulated. That way at least some of the illegality would be curtailed as the city struggles to provide access to clean water for its citizens.

Similar arguments have been made for the legalization of prostitution – an industry the world over that has been immensely susceptible to extortion due to its presence in the nighttime economy. In 2016 Amnesty International formally released a new evidence-based policy calling for the legalization and protection of sex workers, whose role in the informal economy leaves them vulnerable to a whole host of human rights abuses, including extortion, and with limited recourse in the formal justice system (Amnesty International 2016). In the spaces where the formal justice and protection mechanisms fail, the illicit protection–extortion economy can thrive. However, without effective regulation, legalization might reduce the profitability of the protection economy but will not appreciably reduce the harm to the vulnerable and the victims unless those mechanisms are also created, monitored, and enforced.

Conclusion

This chapter identifies extortion as a critical challenge for those concerned with urban security, and argues for a focus by municipalities on extortion as a primary threat, rather than a predicate crime secondary to other concerns such as mafia behaviour, terrorism, or violence. Based upon a set of globally dispersed case studies, the chapter argues for innovative thinking in how to respond to the challenge: less as a criminal justice issue and more as a social phenomenon with deep roots in the local political economy. Effective reduction of levels of extortion will require a framework of responses embedded in a comprehensive, holistic, equitable, and development-orientated approach to urban change. Whole-of-community approaches that shed light on extortion where it exists and mobilize collective momentum for social change have seen notable success.

The chapter reviews the life cycle of an extortion economy, including the conditions required for extortion practices to develop and the typical evolution and entrenchment of the practice. It also highlights when changes to established extortion economies might offer opportunities for intervention to break these economies. Given that it is far easier to prevent extortion economies consolidating than to eradicate them once they exist, the chapter suggests that all municipalities would be well advised to sensitize themselves to the enabling factors and warning signs that indicate an extortion economy is developing.

Disrupting an extortion economy is by no means an easy or straightforward challenge, and there is far too little research being done in this area. But given the painful evidence from multiple continents that extortion has become a social, political, and economic challenge – a death by 1,000 cuts – it urgently and actively needs to be moved higher on the urban security agenda.

Notes

1 Presentation by a senior law enforcement officer from El Salvador at a meeting organized by the Global Initiative against Transnational Organized Crime and the US International Bureau for Narcotics and Law Enforcement (INL), 'Promoting Community Responses to Counter Extortion: Identifying Promising Practices', San Salvador, 28–29 November 2017.

2 Presentation by a senior law enforcement officer from Honduras at a meeting organized by the Global Initiative against Transnational Organized Crime and the US INL, 'Promoting Community Responses to Counter Extortion: Identifying Promising Practices', San Salvador, 28–29 November 2017.

References

Adams, T. M. 2011. 'Chronic Violence and Its Reproduction: Perverse Trends in Social Relations, Citizenship, and Democracy in Latin America.' Washington DC: Woodrow Wilson Center, available at www.wilsoncenter.org/sites/default/files/LAP_111121_chronicviol2011_single_page.pdf (accessed 24 June 2018).

Ahmed, N. 2008. *Water Supply in Karachi.* Karachi and Oxford: Oxford University Press.

Amnesty International. 2016. 'Policy on State Obligations to Respect, Protect and Fulfil the Human Rights of Sex Workers.' London: Amnesty International, 26 May, available at www.amnesty.org/en/documents/pol30/4062/2016/en/ (accessed 24 June 2018).

Anzola, D. 2016. 'Basic Dynamics of Extortion Racketeering', in C. Elsenbroich, D. Anzola, and N. Gilbert (eds) *Social Dimensions of Organised Crime*. Cham: Springer, 26–45.

Anzola, D., M. Neumann, M. Möhring, and K. Troitzsch. 2016. 'National Mafia-Type Organisations: Local Threat, Global Reach', in C. Elsenbroich, D. Anzola, and N. Gilbert (eds) *Social Dimensions of Organised Crime.* Cham: Springer, 9–23.

Ballesteros, C. 2017. 'Mexican Cartels Used Government Data to Kidnap and Extort Avocado Farmers', *Newsweek*, 30 October, available at www.newsweek.com/cartels-kidnap-avocado-farmers-696301 (accessed 24 June 2018).

Beltrán, A. 2017. 'Children and Families Fleeing Violence in Central America', *WOLA*. 21 February, available at www.wola.org/analysis/people-leaving-central-americas-northern-triangle/ (accessed 24 June 2018).

Business Recorder. 2017. 'DG Rangers Claim Sharp Decline in Crimes in Karachi', *Business Recorder*, 25 August, available at https://fp.brecorder.com/2017/08/20170825212257/ (accessed 24 June 2018).

Caballero, J. P. 2016. '"New" Vigilante Groups Keep a Violent Tradition Alive in Michoacán', *InSightCrime*, 13 April, available at www.insightcrime.org/news/analysis/new-vigilante-groups-in-michoacan-mexico-resemble-predecessors/ (accessed 24 June 2018).

Cameron, R. 2009. 'The Single Public Service: The Recentralisation of South African Local Government?', paper presented at 21st World Conference of International Political Science Association, 12–16 July, Santiago.

Chin, K. 1996. *Chinatown Gangs: Extortion, Enterprise, and Ethnicity*. New York: Oxford University Press.

Dixon, B. and L.-M. Johns. 2001. *Gangs, Pagad and the State: Vigilantism and Revenge Violence in the Western Cape*, Violence in Transition Series, Vol. 2. Johannesburg: Centre for the Study of Violence and Reconciliation, available at http://csvr.org.za/docs/gangs/gangspagadstate.pdf (accessed 24 June 2018).

Dugard, J. 2001. *From Low Intensity War to Mafia War: Taxi Violence in South Africa (1987–2000)*, Violence in Transition Series, Vol. 4. Johannesburg: Centre for the Study of Violence and Reconciliation, available at www.csvr.org.za/docs/taxiviolence/fromlowintensity.pdf (accessed 24 June 2018).

Einashe, I. 2016. 'Meet the Migrants Who Are Helping Italians Take on the Sicilian Mob', *International Business Times*, 14 October, available at www.ibtimes.co.uk/meet-migrants-who-are-helping-italians-take-sicilian-mob-1586348 (accessed 24 June 2018).

Fox News. 2017. 'Mexican Cartels Extorting Avocado Producers Raises Price of Guacamole, Report Says', Fox News, 3 November, available at www.foxnews.com/world/2017/11/03/mexican-cartels-extorting-avocado-producers-raises-price-guacamole-report-says.html (accessed 24 June 2018).

Gambetta, D. 1990. 'Mafia: The Price of Distrust', in D. Gambetta (ed.) *Trust: Making and Breaking Cooperative Relations*. Oxford: Blackwell, 158–175.

Gambetta, D. 1993. *The Sicilian Mafia: The Business of Private Protection*. Cambridge, MA: Harvard University Press.

Gambetta, D. and P. Reuter. 1995. 'Conspiracy Among the Many: The Mafia in Legitimate Industries', in G. Fiorentini and S. Peltzman (eds) *The Economics of Organized Crime*. Cambridge: Cambridge University Press, 116–135.

Glenny, M. 2015. *Nemesis: One Man and the Battle for Rio*. London: Vintage Press.

Goga, K., E. Salcedo-Albaran, and C. Goredema. 2014. 'A Network of Violence: Mapping a Criminal Gang Network in Cape Town.' Pretoria: Institute for Security Studies, November, available at https://issafrica.s3.amazonaws.com/site/uploads/Paper271V2.pdf (accessed 1 July 2018).

Hassan, S. R. 2013. 'Pakistan's Commercial Hub Faces Growing Extortion Menace', *Reuters*, 2 July, available at www.reuters.com/article/us-pakistan-extortion/pakistans-commercial-hub-faces-growing-extortion-menace-idUSBRE9601E420130701 (accessed 24 June 2018).

Hobbs, D., P. Hadfield, S. Lister, and S. Winlow. 2003. *Bouncers: Violence and Governance in the Night-Time Economy*. Oxford: Oxford University Press.

Hussain, N. 2016. 'Tracking Order in Seeming Chaos: Understanding the Informal and Violent Political Order in Karachi', input paper to UNODC/University of Cape Town study on Governing Safer Cities in a Globalized World, available on request from the chapter authors.

ICG (International Crisis Group). 2014. 'Policing Urban Violence in Pakistan', 23 January, available at www.crisisgroup.org/asia/south-asia/pakistan/policing-urban-violence-pakistan (accessed 24 June 2018).

ICG (International Crisis Group). 2017a. 'Mafia of the Poor: Gang Violence and Extortion in Central America', 6 April, available at www.crisisgroup.org/latin-america-caribbean/central-america/62-mafia-poor-gang-violence-and-extortion-central-america (accessed 24 June 2018).

ICG (International Crisis Group). 2017b. 'Pakistan: Stoking the Fire in Karachi', 15 February, available at www.crisisgroup.org/asia/south-asia/pakistan/284-pakistan-stoking-fire-karachi (accessed 24 June 2018).

Igarapé. 2017. 'Homicide Monitor: Data for 2017', available at https://homicide.igarape.org.br (accessed 24 June 2018).

INEGI (Instituto Nacional de Estadística y Geografíca). 2014. 'Encuesta Nacional de Victimización y Percepción sobre Seguridad Pùblica (ENVIPE).' Aguascalientes: INEGI, available at www.beta.inegi.org.mx/proyectos/enchogares/regulares/envipe/2014/ (accessed 24 June 2018).

Izadi, E. 2015. 'Driving a Bus is a "Death Sentence" in El Salvador's Capital City', *Washington Post*, 31 July, available at www.washingtonpost.com/news/worldviews/wp/2015/07/31/driving-a-bus-is-a-death-sentence-in-el-salvadors-capital-city/?noredirect=on&utm_term=.94f431bf1773 (accessed 26 October 2017).

Jaitman, L. and I. Torre. 2017. 'A Systematic Approach to Measuring the Costs of Crime in 17 Latin American and Caribbean Countries', in L. Jaitman (ed.) *The Costs of Crime and Violence: New Evidence and Insights in Latin America and the Caribbean*. Washington, DC: IADB, 19–29.

Khan, N. 2010. *Mohajir Militancy in Pakistan: Violence and Transformation in the Karachi Conflict*. New York: Routledge.

Kinosian, S. and A. Albaladejo. 2016. 'El Salvador's Security Strategy in 2016: Change or More Mano Dura?', Latin America Working Group, 29 February, available at http://lawg.org/action-center/lawg-blog/69-general/1599-el-salvadors-security-strategy-in-2016-change-or-more-mano-dura- (accessed 24 June 2018).

Kunbhar, Z. 2016. 'Karachi's Burgeoning Water Mafia', *The Third Pole*, 10 November, available at www.thethirdpole.net/2016/11/10/karachis-burgeoning-water-mafia/ (accessed 24 June 2018).

Martínez, O., E. Lemus, C. Martínez, and D. Sontag. 2016. 'Killers on a Shoestring: Inside the Gangs of El Salvador', *New York Times*, 20 November, available at www.nytimes.com/2016/11/21/world/americas/el-salvador-drugs-gang-ms-13.html?_r=0 (accessed 24 June 2018).

Merelli, A. 2016. 'The Center of Sicily's Biggest City Was Emptied by the Mafia. Now It's Being Reclaimed by Migrants', *Quartz*, 21 June, available at https://qz.com/704320/migrants-are-bringing-back-to-life-palermos-historical-center-which-the-mafia-had-ravaged/ (accessed 24 June 2018).

Newsline Magazine. 2011. 'Extortion: The Top Crime in Karachi', *Newsline Magazine*, June, available at http://newslinemagazine.com/magazine/extortion-the-top-crime-in-karachi/ (accessed 24 June 2018).

O'Toole, M. 2017. 'Micro-Extortion by Gangs is Costing El Salvador Millions of Dollars a Year, $10 at a Time', *Quartz*, 17 June, available at https://qz.com/1008580/micro-extortion-by-gangs-is-costing-el-salvador-4-billion-a-year-10-at-a-time/ (accessed 24 June 2018).

Pauw, J. 2017. *The President's Keepers: Those Keeping Zuma in Power and Out of Prison*. Tafelburg: NB Publishers.

Petrus, T. and I. Kinnes. 2018. 'New Social Bandits? A Comparative Analysis of Gangsterism in the Western and Eastern Cape Provinces of South Africa', *Criminology and Criminal Justice* (January), 1–18.

Puebla On Line. 2017. 'Crimen Organizado Usa Bases de Datos del Gobierno Federal Para Extorsionar y Secuestrar', 29 October, available at www.pueblaonline.com.mx/2017/portal/index.php/seguridad/item/59600-crimen-organizado-usa-bases-de-datos-del-gobierno-federal-para-extorsionar-y-secuestrar#.WfchtVtSzcs&panel1-1 (accessed 24 June 2018).

RT. 2017. 'México, Sacudido por la Fiebre del "Oro Verde"', 25 March, available at https://actualidad.rt.com/actualidad/234163-mexico-pais-oro-verde (accessed 24 June 2018).

Schelling, T. 1984. *Choice and Consequence*. Cambridge, MA: Harvard University Press.

Sciandra, E. 2017. 'Countering Extortion Racketeering: The Italian Experience', Centre for the Study of Democracy, Sofia, available at www.transcrime.it/wp-content/uploads/2017/07/ProjectCEREU-ProjectReader_eng.pdf (accessed 24 June 2018).

Shackle, S. 2015. 'Karachi Vice: Inside the City Torn Apart by Killings, Extortion and Terrorism', *Guardian*, 21 October, available at www.theguardian.com/world/2015/oct/21/karachi-vice-inside-city-riven-by-killings-kidnappings-and-terrorism (accessed 24 June 2018).

Shaw, M. 2016a. '"We Pay, You Pay": Protection Economies, Financial Flows, and Violence', in H. Matfess and M. Miklaucic (eds) *Beyond Convergence: World Without Order*. Washington, DC: National Defense University, 235–250.

Shaw, M. 2016b. 'A Tale of Two Cities: Mafia Control, the Night Time Entertainment Economy and Drug Retail Markets in Johannesburg and Cape Town, 1985–2015', *Police Practice and Research* 17(4), 353–363.

Shaw, M. 2017. *Hitmen for Hire: Exposing South Africa's Underworld*. Cape Town: Jonathan Ball.

Shaw, M. and S. Haysom. 2016. 'Organised Crime in Late Apartheid and the Transition to a New Criminal Order: The Rise and Fall of the Johannesburg "Bouncer Mafia"', *Journal of Southern African Studies* 42(4), 577–594.

Skaperdas, S. (2001). 'The Political Economy of Organised Crime: Providing Protection When the State Does Not', *Economics of Governance* 2(3), 173–202.

Székely, A., G. Andrighetto, and L. Nardin. 2016. 'Social Norms and Extortion Rackets', in C. Elsenbroich, D Anzola, and N. Gilbert (eds) *Social Dimensions of Organised Crime*. Cham: Springer International, 49–64.

Tanoli, Q. 2017. '"Several Milestones Achieved" in Karachi Operation', *Express Tribune*, 30 April, available at https://tribune.com.pk/story/1397316/several-milestones-achieved-karachi-operation/ (accessed 24 June 2018).

Tariq, W. 2015. 'The Importance of Karachi', *Express Tribune*, 12 October, available at https://tribune.com.pk/story/971188/the-importance-of-karachi/ (accessed 24 June 2018).

Thamm, M. 2018. 'Cape Town Mafia: Biggest Underworld Trial to Expose Politically Connected in City's Economy of the Night', *Daily Maverick*, 15 January, available at www.dailymaverick.co.za/article/2018-01-15-cape-town-mafia-biggest-underworld-trial-to-expose-politically-connected-in-citys-economy-of-the-night/#.Wqt39WaZPVo (accessed 26 June 2018).

Tilly, C. 1985. 'War Making and State Making as Organised Crime', in P. Evans, D. Rueschemeyer, and T. Skocpol (eds) *Bringing the State Back In*. Cambridge: Cambridge University Press, 169–191.

Times of Islamabad. 2017. 'Karachi Operation: Crime Rate Drastically Reduced in Metropolitan City', *Times of Islamabad* (English), 5 April, available at https://timesofislamabad.com/karachi-operation-crime-rate-drastically-reduced-metropolitan-city/2017/04/05/ (accessed 24 June 2018).

Transcrime. 2009. 'Study on Extortion Racketeering: The Need for an Instrument to Combat Activities of Organised Crime', University of Trento, available at https://ec.europa.eu/home-affairs/sites/homeaffairs/files/doc_centre/crime/docs/study_on_extortion_racketeering_en.pdf (accessed 24 June 2018).

van der Borgh, C. and W. Savenije 2015. 'De-Securitising and Re-Securitising Gang Politics: The Funes Government and Gangs in El Salvador', *Journal of Latin American Studies* 47(1), 149–176.

Varese, F. 2001. *The Russian Mafia: Private Protection in a New Market Economy*. Oxford: Oxford University Press.

Varese, F. 2013. 'Protection and Extortion', December, available at http://federicovarese.com/sites/default/files/Protection_and_Extortion.pdf (accessed 4 October 2018).

Varese, F. 2014. 'Protection and Extortion', in L. Paoli (ed.) *The Oxford Handbook of Organized Crime*. Oxford: Oxford University Press, 345–358.

Volkov, V. 2002. *Violent Entrepreneurs: The Use of Force in the Making of Russian Capitalism*. Ithaca, NY: Cornell University Press.

von Boemcken, M. 2013. *Between Security Markets and Protection Rackets: Formations of Political Order*. Berlin: Budrich Unipress.

WEF (World Economic Forum). 2017. *The Global Competitiveness Report 2017–2018*. Geneva: WEF.

Yusuf, J. 2013. 'Menace of Extortion', *Dawn*, 4 December, available at www.dawn.com/news/1060326 (accessed 24 June 2018).

Zia, A. 2011. 'The Business of Crime', *Newsline Magazine*, April, available at http://newslinemagazine.com/magazine/the-business-of-crime/ (accessed 24 June 2018).

6 Beyond the usual suspects

Gender considerations at the interface between urban safety and peacebuilding

Jennifer Erin Salahub and Mayssam D. Zaaroura

Introduction

Research on the confluence of urban safety and peacebuilding practice has grown considerably in recent years. The existing literature explores the academic and practical intersections between peacebuilding and urban safety, and provides needed insight and perspectives into important linkages between the two fields of practice and how they support one another.[1] To date, much of this literature has drawn from work on peacebuilding, suggesting that practices such as mediation, trust building, and non-violent conflict resolution used to resolve macro-level armed conflicts are relevant and could be applied to improve situations of chronic violence at the city level. The opposite – if or how more micro-level urban safety theory and practice can support more macro-level peacebuilding efforts – is considerably less explored in this literature and represents an important area for future research; however, this goes beyond the bounds of this chapter.

This new literature is promising, but to date it does not include a systematic analysis of gender. In this chapter we try to identify where and how taking gender as the starting point for analysis (rather than compartmentalizing it or adding it on as one of many layers of analysis) can strengthen the potential of the urban safety–peacebuilding nexus to prevent and reduce violence. The first section briefly presents the working definitions of urban safety and peacebuilding underlying the chapter's analysis. It also shows where and how the existing literature on the confluence of urban safety and peacebuilding and their practices neglect the gendered dimensions of each, notwithstanding robust gender-sensitive literatures in each distinct field and the real-life challenges faced by – and opportunities for – women in cities. The next section identifies the risks of excluding gender and argues for the inclusion of a robust gender analysis that includes men and women, masculinities and femininities in debates about an urban safety–peacebuilding nexus.[2] We conclude with a discussion of entry points and emerging lessons to strengthen gender analysis and its integration into practice, with a view to improving operational effectiveness and making cities safer.

Methodologically, we build on a literature review and draw on our experience and past research on the gendered dimensions of urban safety, security, and peacebuilding. Specifically, we review the series of papers produced by the Technical Working Group (TWG) on the Confluence of Urban Safety and Peacebuilding Practice (see Chapter 1, 4, and 7 in this volume, as well as Grob *et al.* 2016a, 2016b; Parent 2016; Wennmann 2016a, 2016b; Wennmann and Ganson 2016) and several related documents (Jütersonke and Krause 2013; Milliken 2013, 2016). The review assesses the extent to which the papers integrate a gender analysis, and identifies where a gender analysis could be incorporated into existing analyses and potential entry points for addressing gaps.

Urban safety, peacebuilding, and gender

Like many terms in the social sciences, urban safety and peacebuilding are two commonly used terms that lack concrete and agreed meanings. We use a broad interpretation of urban safety that focuses on freedom from fear of harm, and a similarly robust understanding of peacebuilding aimed at establishing a sustained positive peace, following Galtung (1964) and discussed further below. Gender, on the other hand, is very clearly defined in the peacebuilding and other literatures as the socially constructed roles assigned to women and men (see Enloe 1989; Kabeer 1994; Munro 2000; Sweetman 2005; Rehn and Sirleaf 2002; Anderlini 2007; Klot 2007; Väyrynen 2010; Myrttinen *et al.* 2014). But, as Sanam Anderlini (2007: 17) notes in the pointed introduction to her formative book, *Women Building Peace*, 'Too often the terms [gender and women] are conflated and used interchangeably.' Gender is not simply a synonym for women and, indeed, we are concerned with drawing attention to how women and men affect and are affected by violence differently. This includes the role of violent masculinities and oppressive femininities, and how they are constructed and reconstructed before, during, and after conflict (see Meintjes *et al.* 2001; Moser and Clark 2001; Pankhurst 2010; Harders 2011). Likewise, we are concerned with how differential access to and use of power by women and men influences their experiences of violence. In this regard, our efforts take a modest step towards a feminist approach to the topic.

Like women and gender, urban safety and urban security are frequently used interchangeably, further muddying understandings. Moreover, in some languages – Spanish, for instance – there is no distinction and a single word, *seguridad*, is used for both 'safety' and 'security', thus simplifying matters. Those who use the term 'urban safety' have tended to focus on questions of crime prevention, including but not limited to violent crime, and tend to take an urban planning approach to solving them, though this is evolving (ICPC 2016; Assiago 2016; UN-Habitat 2007). Within UN-Habitat's Safer Cities Programme, an early focus on urban crime prevention expanded to a 'multi-dimensional approach to urban safety', including security of tenure and natural disasters (UN-Habitat 2016). UN-Habitat's more recent focus on 'urban security' links safety in cities to the human security agenda and to broader disciplinary approaches to preventing violence (Assiago 2016; UN-Habitat 2007, 2016; Humansecurity-cities.org 2006). But notwithstanding the efforts of UN-Habitat to promote a human security approach in cities, the evolution of practice – such as the Police Pacifying Units in Rio de Janeiro – suggests that speaking of 'urban security' reflects militarized and 'hard security' approaches to urban violence reduction and prevention (Muggah 2012: 55–56).

The focus on urban planning naturally leads to interventions that seek to alter public spaces and infrastructure: urban upgrading, urban renewal, and crime prevention through environmental design (Moser 2004: 11; Muggah 2012: 61–64). Likewise, emphasizing security (human or otherwise) tends to place the state in the centre of the response, with interventions aimed at judicial and police reform, for example (Moser 2004: 11; Humansecurity-cities.org 2006).

From a gendered perspective, both of these tendencies are problematic. Cities are constituted by both public and private spaces in which women and men operate, but which place different rules on them depending on where they are and with whom. Built environment interventions that are 'gender blind' risk exacerbating existing or creating new but separate and different risks for women and men. Expanding our understanding of the spaces in which violence exists opens up new avenues for violence prevention policy and intervention. Particularly when it comes to violence suffered by women, the domestic sphere is just as important as the public sphere (if not more so), and it is sometimes impossible to separate the two (Anwar *et al.* 2016; UN Women 2016b; Dymén and Ceccato 2012; Buss 1997). Similarly, societies place different expectations on a man when it comes to violence depending on whether he is in public or at home (Anwar *et al.* 2016; Slegh *et al.* 2016). Likewise, in peacebuilding processes an often-ignored domestic sphere holds the potential to contribute to demilitarization and prevention of a return to conflict by helping to reorient social valorization of what Enloe (1983: 87) calls 'warrior masculinities', particularly if taken into consideration during disarmament, demobilization, and reintegration processes.

To demonstrate the importance of expanding spaces – both intellectual and physical – of analysis, consider briefly how gender-specific expectations and social norms affect women's and men's safety in Pakistani cities. For women, the domestic sphere is typically safe, providing they follow the rules and fulfil their socially expected roles in the household: cooking, cleaning, and caring. Transgressions often result in violence or threats of violence (Anwar *et al.* 2016). As Anwar *et al.* (ibid.: 93) note, 'In the home, family and relatives are a constituent part of social capital for reducing vulnerability. But they are also a key component in policing the mobility of young women'. The authors further note that women's male relatives will control their access to communications technology and their ability to move about in the city's public spaces. Women are prevented from going to school or work because of concerns about their exposure to violence. When women are in public spaces they are subject to gendered harassment, threats, and physical violence, mainly perpetrated by men. This violence can be catastrophic or mundane. Women face being caught up in ethnic violence that erupts unpredictably, but frequently, at key unavoidable transport hubs. Public transit is equally risky. Women risk being left on the side of the road in unfamiliar parts of the city by unscrupulous bus drivers seeking to maximize profits by creating space for more male passengers. In either case, women's and men's prescribed gender roles come into play to incentivize them to action, to put them at risk, or to constrain their mobility (ibid.: 84–85, 95–96).

Working and thinking only at the level of the state further obscure the necessity of a gender analysis, as this approach moves away from the people who are responsible for urban violence or armed conflict and are its victims. Bringing an analysis closer to the level of the people who experience insecurity and safety allows us to ask 'safety for whom?' and 'insecurity for whom?' Expanding the focus beyond state-centric approaches can strengthen gender analyses and understandings of what it means to be insecure or safe in a city. It also creates space for actors other than national governments – such as municipalities, women's organizations, other civil society groups, and the private sector – to be the central figures in any violence prevention or reduction activities. This is not to suggest that it absolves the different state authorities of their responsibility to provide safety or security in cities, but rather suggests that gender analyses become more obvious when people are the unit of analysis rather than the state.

Peacebuilding – especially the approaches prevailing in the United Nations (UN) system – has struggled with this shift from the state to the individual as a unit of analysis. The transformative power of individuals and specific groups – such as women and women's movements – in peacebuilding is restricted at both a structural and a practical level. Structurally, the institutions and systems of peacebuilding, like all patriarchal structures, systematically exclude most women. This exclusion is carried through to practice, where women are ignored and their agency is constrained (Anderlini 2007). The same holds true for peacebuilding approaches that incorporate analyses of masculinities (Sudhakar and Kuehnast 2011; Saferworld 2014). Notwithstanding some positive efforts on the part of researchers and civil society actors in the last decade, including those referenced in this chapter, progress within the UN system – the key peacebuilding actor – has been slow and insufficient. Part of this problem relates to continued confusion and divisions over what peacebuilding means in theory and in practice, and where gender fits as a conceptual and implementable dimension within this field (Pankhurst 2010; El-Bushra 2012). A wide range of peacebuilding definitions and approaches exist (Chetail and Jütersonke 2015), which – perhaps counterintuitively – provides much operational space for peacebuilding professionals, but also can contribute to ambiguities about roles and responsibilities of individuals, groups, institutions, and state authorities.

Galtung (1964) originally used the term 'peacebuilding' to be more encompassing in its activities in comparison to previous conceptualizations. He coined the now-common terminology of 'negative' and 'positive' peace: the former describing peace in terms of the absence of violence, the latter as the prevalence of just, harmonious, and equal relationships free from cultural and structural violence. After the end of the Cold War, peacebuilding became increasingly associated with UN peacebuilding efforts as articulated by then Secretary-General Boutros Boutros-Ghali in his Agenda for Peace (UNSG 1992). The

Agenda for Peace brought peacebuilding terminology more 'systematically into UN vocabulary' (Wennmann and Ganson 2016: 2), although it should be noted that it referred to peacebuilding as 'post-conflict peacebuilding', thereby limiting the application of the term to activities taking place when conflict has ended, or is coming to an end.

Over the past decade the range of actors involved in peacebuilding practice has grown and the bounds of when and how peace is built and by whom has expanded the field of practice beyond the Agenda for Peace definition. With respect to post-conflict peacebuilding, some observers emphasized that societies should have a chance to 'build back better' (Goetz and Jenkins 2015: 212). UN Security Council Resolution 2282 (UNSC 2016) recognized this evolution and marked the 10-year review of the UN peacebuilding architecture 'that repositions peacebuilding as activity across all stages of conflict and as a priority for the entire UN system' (Wennmann and Ganson 2016: 2). Importantly, the resolution emphasizes 'sustaining peace' through an inclusive and normative process (ibid.: 2–3).

In this way, the concept of 'peacebuilding' seems to have returned to Galtung's 'positive' peace. Such a broader conceptualization can be applied to urban safety as part of a comprehensive strategy to ensure sustainable and equitable peace, creating space to discuss normative visions of an ideal society that includes the promotion and nurturing of new fields of citizenship and political participation for all. Nevertheless, the details of how to achieve a positive peace are rarely discussed in the peacebuilding literature, creating a considerable challenge for implementation. What is more, ending hostilities – achieving a negative peace – is often prioritized, sacrificing the gendered analysis that would contribute to a more lasting peace and, as Pankhurst (2010: 156) argues, 'leaving major causes of violence and war unresolved – including not only economic inequalities, but also major social divisions and the social celebration of violent masculinities'.

A focus solely on 'post-conflict' peacebuilding, and in particular one that disregards a more nuanced gendered analysis including both women and men, is too narrow to build a positive peace. This is because post-conflict peacebuilding frequently concentrates on the transformation of state authority and institutions, reinforcing and hardening a state-centric order without addressing structural sources of conflict, such as social hierarchies (including gendered hierarchies), economic exclusion, or stigmatization. Such post-conflict peacebuilding – frequently supported by international donors and the UN – can favour stabilization of a turbulent situation over gender-sensitive conflict transformation, or even a return to the status quo, reinstating pre-conflict institutions and unequal gender relations if this means a 'more stable environment' measured through instruments that themselves reflect an unacknowledged gender bias. For example, a more stable environment could be defined by a cessation of hostilities – an important objective, but one which usually overlooks the gendered dimensions of armed conflict and violence, including low female participation in peace negotiations and the disproportionate burden of domestic violence borne by women. In other words, while the fighting may have stopped, women (and many men) are unable to ensure their needs are met in peace negotiations or to be safe in their own homes. We contend that these unequal gender relations – focused on access to and capacity to use social, political, and economic power, and rooted in oppressive patriarchal structures – are key contributors to violence and armed conflict.

For peacebuilding practice to share lessons with urban safety, it must incorporate a robust gendered understanding and analysis of how identities are constructed through societal relations between and among women, men, boys, and girls, and how those identities contribute to creating and recreating unequal power relations. This 'gender-relational approach' (Myrttinen *et al.* 2014) should be the starting point of any peacebuilding policy or programme aimed at a sustainable and positive peace, as it allows a comprehensive assessment of roles and relations in a particular context. Such an approach provides much more than simple sex-disaggregated data collection and statistical analysis, although these are useful in identifying the harmful consequences of gender biases and the scale and scope of problems (see, for example, Geneva Declaration Secretariat 2011, 2015). It requires a more engaged analysis looking at reasons *why* women and men experience violence differently; it leads to a nuanced understanding of a society's potential for violence and conflict, and opens the door to identifying a broader range of interventions, actors with whom to work, and targeted policies that are more effective. In practice, this can become clear in programming that seeks to change social and cultural norms to prevent violence before it happens by 'creat[ing] cultural shifts in what is acceptable and unacceptable behaviour, and in promoting norms supportive of healthy, non-violent and gender equitable relationships' (WHO 2014: 28).

Understanding the risks of excluding gendered analysis

The papers published under the auspices of the TWG on the Confluence of Urban Safety and Peacebuilding identify how peacebuilding and urban safety practice can collaboratively make cities safer. However, they do not include a gender analysis, and ignore gender at both systemic and operational levels. They fail to recognize the different experiences of violence by different groups of men and women in meaningful ways. The papers do not address the gendered power structures and social norms that shape and contribute to men and women's different experiences of violence. Most of them assume (incorrectly) that policies and programmes affect women and men equally, while at the same time ignoring the heterogeneity of these two groups. In so doing, they neglect the agency of women and men in their analyses. This section briefly reviews the existing stock of analyses on the confluence of urban safety and peacebuilding practice with respect to their shortcomings on gender. It also raises opportunities to integrate gender more systematically and meaningfully into the analysis at the confluence of peacebuilding and urban safety practice.

Overall, the TWG papers are based on a conceptualization of peacebuilding that allows space for a gender narrative and is consistent with the notion of positive peace, but then fails to acknowledge and integrate gendered analyses systematically. Content analysis can help reveal gendered differences: among the papers, Parent (2016) includes the largest number of references to women; however, all but one of those references are linked to children, including ‘women and youth’ or ‘women and minors’, a persistent trope that limits women’s independent agency and infantilizes them. Grob *et al.* (2016a, 2016b), Wennmann (2016a), Wennmann and Ganson (2016), and a number of the contributions to this volume can also be faulted for their lack of engagement with gender. Contributions by Milliken (2013, 2016) that document proceedings at two meetings do incorporate a meaningful discussion of the specific safety challenges faced by women, such as violence against women and girls. Additionally, Milliken (2016) begins to incorporate a gendered analysis, reflecting an encouraging evolution of the discourse.

Engagement with the male side of gender – men and masculinities – reflects the paucity of work on these issues more broadly. Only two of the core TWG papers on the confluence of urban safety and peacebuilding practice (Grob *et al.* 2016b; van den Eertwegh 2016) use the word ‘men’, one instance of which is in a direct quote from a supporting document. Masculinities do not figure in any of the analyses of the TWG despite a critical mass of research on masculinities more broadly, and recent studies of masculinity and peacebuilding in particular, that stands to strengthen both analysis and practice on the confluence of urban safety and peacebuilding (see, for example, Connell and Messerschmidt 2005; Harland and McCready 2014; Pankhurst 2016).

This brief analysis suggests that current work about the confluence of urban safety and peacebuilding practice does not account for its gendered dimensions in any systematic way. This lack of a gender analysis is problematic for peacebuilding and urban safety practice, for at least three reasons.

First, the absence of a gender analysis means that peacebuilding and urban safety practice is missing practical opportunities to be more relevant and reduce risk. For instance, gender analyses would allow the peacebuilding and urban safety practice community to overcome the strong focus on large-scale violence and turbulent political transitions that peacebuilding and many planning-based urban safety interventions tend to target and include the more mundane, less physically harmful violence that – arguably – more people experience on a daily basis in their households (Anwar *et al.* 2016). Not surprisingly, much of that violence is perpetrated by men against women.[3]

Second, the absence of a gender analysis marginalizes the different security needs of women, some men, girls, and boys, and tends to cater disproportionately to the security needs of wealthy, powerful men. Without a gender analysis, urban safety and peacebuilding programmes fail to live up to human rights standards and exhibit reduced operational effectiveness. Excluding women from traditionally male-dominated discussions of security and safety undermines human rights and goes against the leading policy guidelines of ensuring participatory processes in transitions to democracy (Nilsson 2012; Paffenholz *et al.* 2016). Failing to take into account the differential security needs of women,

men, girls, and boys and excluding women from serving in security services and their perspectives from security sector reform (SSR) agendas lead to one-size-fits-all service delivery that cannot meet the needs of the population it serves (Salahub and Nerland 2010).

Finally, the absence of a gendered analysis serves to reinforce existing exclusive and exclusionary power structures. From a peacebuilding perspective this is counterintuitive, as it is the very social structures that facilitate armed violence and conflict that should be transformed to achieve a positive peace. Failing to incorporate a meaningful gendered analysis – to take gender as a starting point rather than compartmentalizing or superficially 'mainstreaming' it[4] – means that, like the TWG papers, peacebuilding and urban safety practice falls into the same trap that it purports to overcome: that of aiming for a hollow, negative peace rather than a positive peace characterized by 'harmony associated with mature relationships' (Wennmann and Ganson 2016: 2). This applies equally to relations between men and women, not least as they are the constitutive actors of any peaceful and inclusive society. Without addressing both the physical and structural violence rooted in oppressive, unequal, and discriminatory social norms, power relations, and roles assigned to men and women – violent masculinities and oppressive femininities – a truly positive peace cannot be achieved, whether between warring nations or rival urban gangs.

In sum, continuing to work within a post-conflict, negative-peace paradigm when approaching the nexus of peacebuilding and urban safety poses considerable risk for importing the drawbacks, if not some of the failings, of peacebuilding into precarious urban contexts. Such an approach also risks exacerbating the pernicious gender dynamics already at play in violent urban settings and serves to prop up the patriarchal social systems that contribute to urban violence rather than realizing the *transformative* potential of peacebuilding and urban safety programmes in cities.

Integrating gender into the confluence of peacebuilding and urban safety practice

Over the last two decades many international and grassroots programmes and initiatives on increasing urban safety have demonstrated greater awareness of the importance of gender, focusing on women's involvement and women's roles to work towards equality. They have recognized the importance of integrating into programming women's different experiences of conflict and violence. Practice at the grassroots level shows notable promise. Organizations such as Women in Cities International, Jagori, the Huairou Commission, Territorios Seguros, and others ranging from grassroots civil society organizations to those working internationally are prioritizing women's security needs in urban spaces (WICI 2012; O'Leary and Viswanath 2011; Salazar Sanchez and Pérez Sáinz 2015).

The trend towards greater gender inclusivity has been even more visible in the research and advocacy supported through the women, peace, and security movement. For over a decade the UN has nominally led intergovernmental efforts in this regard since its Security Council passed Resolution 1325 (UNSC 2000) on women, peace, and security. The resolution represented a landmark moment based on decades of engagement and advocacy; but since its adoption, implementation has lagged behind expectations and gender has yet to be robustly and meaningfully integrated into the activities of the UN and its member states, with the potential exception of policy and awareness-raising efforts to reduce violence against women and girls.

Since 2000 the Security Council has passed several 'sister resolutions' to Resolution 1325, including Resolution 2122 (UNSC 2013) aiming to strengthen women's role in all stages of conflict prevention, and Resolution 2242 (UNSC 2015), marking the 15th anniversary of 1325. Attempting to overcome the gender-as-synonym-for-women obstacle, these resolutions pushed the onus on to international and regional organizations and UN member states to move beyond their narrow view of sexual violence, translate commitments into action on the ground, and focus on strengthening women's leadership and engagement in the policy-level peacebuilding agenda. This means more focused attention on and asking questions about root causes: 'gendered societal power relations and identities' that are seldom addressed (Myrttinen *et al.* 2014). More notably, and we discuss this further below, Resolution 2242 recognized the role that women's organizations can play in meaningfully translating commitment to action (Hans and Rajagopalan 2016).

These practical efforts to integrate gender are reflected in increased scholarly interest in gender and peacebuilding, which is easily demonstrated by a simple online search. For example, a Google Scholar search on 31 January 2017 for 'gender and peacebuilding' yielded more than 13,000 results in indexed publications since 2013, excluding citations, and close to 29,000 overall. A parallel search for 'women and peacebuilding' yielded 35,400 results overall, of which 14,700 are from publications since 2013. When the latter search is run excluding the word gender – i.e. for publications that speak to women and peacebuilding but do not discuss gender – there are 17,800 overall, but only 6,800 since 2013. What this blunt instrument suggests is an evolution of the discourse around gender and peacebuilding similar to the evolution of research and practice on gender and international development, which began as 'women *in* development', then became 'women *and* development', and has settled on 'gender and development' (Rathgeber 1990). Myrttinen *et al.* (2014: 10–15) describe the peacebuilding version of the 'WID–WAD–GAD' evolution in its shift from a 'women, peace, and security' agenda to a 'gender and security' agenda that rejects the use of gender as a synonym for women, challenges the essentialization of men as perpetrators and women as victims, and highlights the importance of understanding the role that different masculinities play in creating violence and peace. Much of this literature is focused on policy and practice, as Anderlini (2007), Klot (2007), Myrttinen *et al.* (2014), Sweetman (2005), and others suggest.

The examples above of how women are regularly being incorporated into urban safety practice and, separately, peacebuilding practice, are encouraging steps, as are emerging efforts to unpack where and how masculinities fit in. Our challenge is to consider how to bring them together and what this means for practice at the nexus of peacebuilding and urban safety. There are several points to be made in this regard. First, there is a transformative potential in both fields that is unrealized. In peacebuilding, this has been long recognized through the 'positive peace' agenda, whereas in urban issues it is relatively new (Moser 2016). Bringing the two fields together creates a new opportunity to shift paradigms and realize the potential to transform (gendered) power relations and build more equal spaces and societies.

In her groundbreaking work on this, Moser (ibid.: 6) highlights six areas of gendered urban interventions that show transformative impacts. Of these, issues of land, informal settlement upgrading, urban safety in public spaces, and water and sanitation offer, in our opinion, the clearest opportunities for peacebuilding to have a positive influence and for peacebuilding tools and strategies to be usefully adapted and applied. In each of these areas, gendered conflicts exist and can spill over into violence in both public and private spaces.

For example, in Ahmedabad and Guwahati, India, low-income populations displaced to allow for urban development projects live in poorly planned settlements where access to basic services, such as water, is limited (Mahadevia and Desai 2019 forthcoming). At the same time, provision of some basic services has been captured by private rent-seeking interests (ibid.) This situation combined with other stressors puts people in conflict, but that conflict is gendered because women are largely responsible for water collection. Using peacebuilding tools such as trust building could play an important role in defusing immediate situations of (potential) violence between individuals and helping to restore faith in state institutions to fulfil their responsibilities, in conjunction with institutional reforms. Similar opportunities exist in gendered conflicts and violence related to land and who can own or inherit it. Likewise, peacebuilding approaches can improve urban upgrading processes that do not take into account the need for sufficient and meaningful community consultation (Barolsky *et al.* 2016) or gendered uses of space.

In considering recommendations for how urban safety and peacebuilding can better integrate gender as they come together, we are guided by three factors: drawing on the separate track records on gender in each field; considering the manner in which the two fields are coming together – i.e. that peacebuilding tools seem, for the moment, to be more relevant to urban safety than vice versa; and being conscious of the need for pragmatic approaches in the short term while more transformative approaches are developed over the longer term. In this light, we offer the following entry points to ensure that gender is central to the confluence of these two streams of work while at the same time seeking to shift the direction of the new joined stream towards a more transformational paradigm shift.

Women's urban safety and security

Recognizing that urban security needs are different for men and women is the first step in making cities safer for both women and men. It is widely known that men are disproportionately the perpetrators *and* the victims of homicide in many informal urban settlements worldwide, while women suffer more domestic, non-lethal violence and are overwhelmingly the victims of sexual violence (Muggah 2012: 30). Female perpetrators of domestic violence are rare and tend to use violence in self-defence, often in the context of existing abusive relationships (Swan *et al.* 2008). Public violence by women is equally rare, though evidence exists to show their role in encouraging men to violence (Baranyi and Powell 2005; Small Arms Survey 2012), and as spectators at and participants in fights and vigilante violence (IDRC 2017). Sex-disaggregated data are routinely collected by national and international bodies, and can be analysed at the level of the city and, in some cases, smaller areas. Incorporating these gender-sensitive data and associated analyses in future urban safety and peacebuilding programmes would strengthen policy and practice.

Understanding how women experience life in cities differently is crucial to making urban spaces safe for them. For example, 92 per cent of women in New Delhi reported having experienced sexual violence in a public space at some point in their life, according to a 2012 UN Women baseline study (cited in UN Women Nd: 2). Women in Ahmedabad, India, frequently use public transit to move about the city, but this activity can be constrained by their fear of sexual harassment and touching on crowded buses (Mahadevia *et al.* 2017: 2, 2016: 1–2). A lack of reliable public infrastructure also constrains women's full access to urban spaces and the activities they take part in. Poor lighting, uneven road services, and unreliable solid waste management all contribute to creating an environment that is hostile to women and adds to their insecurity (Mahadevia *et al.* 2016: 2–3).

In considering a gendered perspective, it remains imperative to be aware of women's heterogeneity when addressing their security. The examples above, from India, speak to the experiences of mainly poor women who lack the means to pay for safer transport and live in more affluent parts of the city. Nevertheless, women in tonier suburbs may face the parallel challenge of negotiating what Dymén and Ceccato (2012: 314) call 'being *in* control versus being *under* control' (emphasis added). By this, they aim to highlight the difference between women's agency in the face of unsafe urban spaces and the imperative that they participate actively and meaningfully in the planning process to create safer spaces, and the surveillance (in public and private spaces) and social control – through gender and other social norms – that restrict their agency, power, and empowerment (ibid.: 325–328).

Work at the confluence of peacebuilding and urban safety practice should enable participatory approaches to identify the different needs of different groups of women and address the tensions between them, rather than focusing on 'women' as a single homogeneous group. Interventions in urban spaces plagued by violence that are based on peacebuilding tools should ensure that women's safety needs are taken into account in any negotiated settlement that relates to how public space is to be managed. At the same time, consideration must be given and appropriate mitigating or remedial action taken to ensure that reductions in public violence or changes in public infrastructure do not have negative cascade effects for women in private spaces. Learning from peacebuilding, women must be active participants in any negotiated settlement. Such an approach is also helpful as it focuses on women's agency and moves away from identifying them solely as victims without agency or voice, and sharing a common set of aims and experiences – or worse, as Pankhurst (2010: 162) puts it, 'as genderless members of other groups'. The same is true, of course, for men.

Masculinities

Recent research confirms that social expectations of what it means to be a man place unrealistic expectations on men with respect to income, consumer goods, and power over women. This phenomenon is frequently exacerbated by urbanization, and fosters crime and violence (Taylor *et al.* 2016; Slegh *et al.* 2016). These violent masculinities are intimately linked with gang cultures and drug trafficking – and police responses to them – in many urban centres globally. In Rio de Janeiro, for example, participants in Taylor *et al.*'s (2016: 47) study speak of 'a "militarized" childhood with daily, widespread exposure to guns and weaponry' wielded and used by drug lords and police as they battle for control of the *favela*. Exposure to this and multiple other forms of public and private violence contribute to making young men in the city's poorest neighbourhoods use violence themselves as they grow up (ibid.: 46–52).

Several TWG papers (Grob *et al.* 2016a; van den Eertwegh 2016; Wennmann 2016b; Wennmann and Ganson 2016) point to the potential of mediation and negotiation – tools that have had success in ending civil wars – to help prevent violence in territories controlled by urban non-state armed groups: gangs, militias, and organized criminals. We contend that incorporating an awareness and analysis of masculinities into those processes would increase their likelihood of success because of the degree to which the violence is tied up in 'being a man'.

Indeed, the field of gender and peacebuilding offers insights into masculinities that are relevant for urban safety efforts. Handrahan (2004: 433) argues that post-conflict losses of gender equality gains made during conflict are the result of the international community imposing its hegemonic masculinity. She writes that while war breaks down patriarchal structures, this gender liberation is short-lived as the (inter)national patriarchy reasserts itself:

> The loosening of rigid and patriarchal gender schemas is closed not by the national male leadership, but by the male international development community, whose own sense of patriarchy-as-normal is quite intact.... This may be evident in the aggressive refusal by the international development community, as a whole, to seriously consider gender issues in post-conflict reconstruction.... The lack of 'gender mainstreaming' seems not to result from an inability on the part of the international community to know better ... but rather suggests its inability to consider its own patriarchy and the damage this does within international development paradigms.
>
> (Ibid.: 436)

Gender-sensitive security sector reform

Grob *et al.* (2016a: 7–8) highlight that SSR is a key institutional entry point for building peace at the city level. Integrating capacity building at a local level and bringing community participation into any security sector training scheme is a laudable endeavour that addresses important gaps in urban safety, particularly in developing countries and many low- and middle-income neighbourhoods (Owusu 2016). Recent research in South Africa and Brazil shows how robust community engagement – beyond consultation – is crucial to ensuring that violence prevention efforts are sustainable and do no harm (Barolsky *et al.* 2016). Yet much opportunity exists to strengthen the focus on the gendered nature of SSR when considering its application to improve urban safety, drawing on the wealth of gender and SSR literature that demonstrates both the principled and operational needs of bringing women into security and defence services (Bastick and Valasek 2008; Eriksson Baaz and Utas 2012) and the strong emphasis on the need to change the culture of violence that imbues security institutions.

Enloe (1993) argues strongly for considering the impact of masculinities on the security sector, deliberately shedding light on how 'manliness' is socially constructed within the military during war and through the police and other armed state institutions, and how this affects the sustainability of post-conflict peacebuilding – and, by extension, urban safety. Writing in the post-Cold War era, at a time when many governments were forming or radically reforming their security institutions and making decisions on the culture that will shape them, Enloe highlighted an uncomfortable truth with regards to demilitarization after war:

> Just how a man (or an adolescent boy) has experienced militarization and how willingly he sheds the habit and expectations of militarization will rebound on the women he returns to. His new definition of his masculinity … will be played out in his family life, in his interactions with women workmates, and in his exchanges with women who are perfect strangers.
>
> (Ibid.: 26)

In this light, building peace through SSR from a gender perspective goes beyond simply increasing the number of women in security institutions, to making operational decisions about discipline, acknowledging homosexuality, and challenging traditional (warrior) masculine authorities and power structures in these institutions that do not provide a safe space and are in fact discouraging women from wanting to join them. It is important to be aware that changing these entrenched norms and processes can take years, and military institutions can be particularly resistant to change in this regard. Even the military forces of developed countries face considerable challenges in incorporating women and homosexuals into their corps in ways that ensure members' personal safety vis-à-vis their colleagues so that their contributions to operational effectiveness can be fully maximized. In the United States the 2010 Act to repeal the 'don't ask, don't tell' policy on homosexuality only *begins* the process administratively and does not place an outright ban on discrimination on the basis of sexual orientation in the military (Arkles and Gehi 2016). It will take years for that starting point to trickle down into practice, and decades longer, if ever, for the culture of acceptance to become the new norm.

By using a feminist lens that unpacks gender (rather than simply using it as a synonym for women), we are able to recognize the role that security institutions in particular play in creating violent masculinities and begin putting in place policies and programmes that counteract this practice. We can then start to look at how more *gender transformative* tools – which some gender equality advocates have used successfully to gain access to post-conflict planning processes, for example (Goetz and Jenkins 2015: 212) – can be applied to recast power structures that oppress both women and men. Enloe (1993: 26) stated, 'It can take years to demilitarize a society. Masculinity and femininity will be among the political territories where the struggles for demilitarization will have to be played out.' We are still struggling with this issue today – whether in theory or in practice – but the opportunity to move forward by bringing peacebuilding and urban safety together is ready to be seized.

Women in decision-making

Women continue to be excluded from many peace negotiations despite evidence that their early and sustained involvement leads to more inclusive and sustainable settlements (UN Women 2016a). At the same time, women bear a disproportionate burden of both armed conflict and many types of urban violence. Failing to take their experiences into account or include them and their voices meaningfully in the design, development, and implementation of any formal or informal peacebuilding or violence prevention intervention creates a serious risk of failure and, potentially, of causing more harm. This failure leaves room for relapse into conflict (Rosenthal 2015: 24). Turshen (1998: 20) argues that 'war … destroys the patriarchal structures of society that confine and degrade women. In the very breakdown of morals, traditions, customs, and community, war also opens up new beginnings.' Yet gender and the objectives of positive peace are rarely discussed among those involved in peace settlements, largely because women are excluded from negotiations.

Recent peace negotiations on Syria and Yemen have set hopeful examples about how to ensure participation of women and women's groups in peace processes. In Yemen the conflict that followed the 2011 revolution has shifted the national government and international community's attention from poverty reduction to stabilization efforts. Shakir's (2015) study assessing the status of Yemeni women's involvement in the country's peacebuilding efforts shows that leading women's advocacy groups have made significant strides in ensuring Yemeni women's voices are heard. In the National Dialogue Conference women represented 30 per cent of participants and actively contributed to the dialogue. However, there remains a large gap in creating a framework for Yemeni women's *meaningful* involvement in conflict resolution and peacebuilding, as Shakir notes:

> Women's meaningful participation in transition processes requires women's representation in terms of numbers and also in terms of gender perspectives (goals and plans) in all aspects of conflict resolution and peacebuilding processes, including post-conflict recovery, political participation in conflict resolution, mediation, security sector reform and transitional justice, as well as state reconstruction. Unfortunately, the research did not identify adequate activities for women's participation amongst these different opportunities and sectors, except for their participation in the NDC [National Dialogue Conference] process at national and community level.
>
> (Shakir 2015: 5)

The situation is similar in Syria. Syrian women's advocacy groups that had been sidelined in the first rounds of failed talks were asked to participate in consultations with the UN special envoy overseeing the peace talks. Early 2016 witnessed the formation of the first ever Syrian Women's Advisory Board – a group of 12 Syrian women from different professional, political, and religious backgrounds – to provide insights into the peace talks (Lock 2016).

While acknowledging the momentous opportunities this represents for Syrian women, it should be noted that the Syrian advisory board, which is composed of civil society actors, did not participate in the decision-making within the formal negotiation process but found alternative methods of making their voices heard, such as through the media (see, for example, Jouejati 2017). Civil society writ large – and in particular women's advocacy groups – are usually excluded from formal peace talks due to the perception that 'introducing new actors could destabilise an already unpredictable process' (Lock 2016: np).

Overall, research on inclusive transition processes suggests that this perception is unfounded, and in fact strongly suggests that the presence of women at formal peace negotiations brings new perspectives to the table that can 'produce positive gains for women and shape a more sustainable and meaningful peace' (ibid.). Research indicates that peace talks are 60 per cent less likely to fail if they include the participation of civil society and women's groups (Nilsson 2012). Nonetheless, despite a general consensus among international mediators stipulating the inclusion and participation of women, in Mali, Syria, Yemen, and other similar contexts women continue to face serious challenges as a result of unfavourable social and political conditions. This is aptly summed up by Tuija Talvitie, executive director of Finland's Crisis Management Institute, in the foreword to Shakir (2015: 3):

> I am convinced that women's greater role in conflict resolution and finding a durable, peaceful solution to the current crisis are mutually reinforcing, rather than competing, priorities. This is exactly the moment to ensure the whole potential of the Yemeni society is utilised for peace and stability of the country.

When considering these peacebuilding efforts in their potential to support urban safety, two key points arise: first, the coming together of two potentially transformational agendas, as discussed elsewhere; and second, the enabling environment of cities where women already serve in many decision-making positions in both city government and civil society. While globally comparable data are lacking, United Cities and Local Governments estimates that approximately 20 per cent of city councillors and 5 per cent of mayors are women (Sáiz and Lasagabaster 2017). Considering the well-known barriers to women's entry into politics, this is a strong base from which to build, and experiences from peacebuilding and the women, peace, and security agenda can further reinforce it. Women leaders in faith-based organizations, civil society, and women's and feminist organizations are also important resources to leverage to make cities safer.

Making gender political

Hale (2001) and Meintjes *et al.* (2001) have questioned why women's gains during conflict fail to be reflected in post-conflict, or post-revolutionary, societies. They identify, among other reasons, a lack of sufficient organization during conflict; women remaining in caretaking roles within a patriarchal order despite unprecedented social upheaval; the detrimental role of international organizations that impose their own internal patriarchal structures and ideologies on local populations and systems; and patriarchal society's devaluing of women's activism. In cases where unequal power dynamics play out in the post-conflict period, the push for gender equality, including recognition of women's agency during conflict, is put aside as secondary to the rebuilding of a nation and the institutional and structural celebration of violent masculinities that accompanies post-war state building, thus emphasizing the state-centric nature of much of peacebuilding policy noted above. The same can be true of cities emerging from urban violence that is highly gendered, such as hyper-masculinized gang violence in cities of Central America.

Gender equality in the post-conflict period is often considered secondary to nation building and reinstating state security at all costs. The term 'gender' is politicized by ruling elites and conflated to mean women's issues, hence limiting the conversation to women as victims and any subsequent state or international action to preventing sexual violence against women and girls. Transformative gender relations fall under the rubric of an inclusive human rights agenda that involves an overhaul of societal relations and state institutions – at the risk of temporarily prolonging conflict and insecurity – and acts as a long-term benefit to women, men, boys, and girls.

Key questions for policymakers and practitioners to explore the gendered political aspects of peacebuilding in urban contexts include *why* women continue to be politicized and marginalized, and have their bodies used as a political tool during and after conflict/violence to uphold patriarchal political structures, and *why* peacebuilding efforts continue to fail in putting political pressure for a human-rights-based approach vis-à-vis gender relations on governments and security institutions – the very same authorities that tend to reinforce patriarchal stereotyping of men as violent and women (activists) as naïve (Enloe 2013: 78). Considering Moser's (2016) analysis that supports the need for collective action and partnerships between government and non-government institutions to bring about transformative change in urban spaces, this latter concern is particularly challenging for realizing the transformational potential of the urban safety–peacebuilding nexus. Analysing the Arab Spring, Amar shows an example of this in post-Mubarak Egypt where state elites play a 'masculinities game' (quoted in Enloe 2013: 78), using essentialized constructs of Egyptian men from a lower class or different ethnicity as 'naturally' posing a risk, and women as victims needing protection, to shore up patriarchal nationalist agendas, state (security) structures, and the use of force.

Dismissive gender stereotypes play a significant role during political transitions, when advocacy groups transform violence against women into a political issue (ibid.: 79). Comments Enloe (ibid.), '"Not now, later" together with "Don't divide the movement", are words of advice that women activists have heard repeatedly – in Vietnam, in France, in Poland, in Algeria, in Eritrea, in Nicaragua, in Mexico, in Iraq and in Chile.' We would add to this list women's recent experiences in Libya during and following Gadhafi's regime, and Egypt during and following Mubarak. Just prior to Gadhafi's death, a delegation of women human rights activists from Libya asked to be included in a high-level UN meeting in New York designed to support the Libyan National Transitional Council (NTC). Fearing that the women might criticize the NTC, UN member state missions said that the women did not 'represent a cross-section of Libyan political life' (Goetz and Jenkins 2015: 219).

For Libyan women to overcome numerous obstacles and lack of resources and build a women's network that could represent a progressive and transformative change for the country on the international peacebuilding stage was a significant feat. Yet the international community's failure in Egypt and Libya to push for more gender-transformative peace settlements has severely impacted women – and men – in these countries, as Daesh[5] and other armed groups and militias use vicious and gratuitous violence against women as a weapon of war in the Middle East and North Africa. In the case of Libya, a month after the fall of Gadhafi's military NTC forces had taken Tripoli and declared their intention to 'annul laws that … violated the tenets of Islam' (ibid.: 220), effectively perpetuating a series of decisions that continue to erode Libyan women's rights and freedoms.

Interviews conducted by Mayssam Zaaroura with women activists in 2013[6] in Egypt and Indonesia document how women in post-Suharto Indonesia had similar experiences, with marginally more success due to a unified women's movement. Indonesia's post-Suharto experience offers valuable insights on the importance of women's organizations in bringing a gendered perspective to urban violence reduction and peacebuilding efforts. Almost two decades after Suharto's fall, increasing religiosity across the country is threatening many gains made by the Indonesian women's movement in 1998, when there was a significant push during and after the overthrow of Suharto's dictatorship. The

Indonesian women's movement – the Gerwani – had been demonized and decimated in the Suharto era for pushing gender norms and becoming politically engaged in the failed 1965 coup attempt. Prominent activists from the Indonesian women's movement interviewed recalled that during the revolts that orchestrated Suharto's fall, the women's movement (composed of feminists from various organizations across the country) put aside significant differences in how they saw gender equality and worked as a unified group to ensure passage of a law requiring a 30 per cent quota of women candidates in elections. They also insisted on the creation of a commission on violence against women at the same time as the human rights' commission, 'literally starting out in the small garden shed behind the newly instated official Human Rights' Commission'.[7]

At that time in Indonesia, merely having the word 'anti' in any title or name – such as anti-domestic violence – was frowned upon; for that matter, the term 'women' was not welcomed much either. However the women's movement built ties with then president-elect B. J. Habibie and made sure these concessions were made. Today, Indonesian activists are fighting for gender mainstreaming in budgetary allocations on a local level to counter religious fundamentalism, building the capacity of women activists to counter the frequent and baseless 'there are no qualified women' refrain, and trying to ensure their presence on electoral commissions and ballots. Crucial components of this success, as recounted by activist women, were having a unified movement and a sympathetic leader, and refusing to wait for state-building processes to take over.

In contrast, the Arab Spring in Egypt saw a divided women's movement that was forcibly pushed out of public squares by sexual assault and state-led intimidation. State security in Cairo was unable to protect women, and in fact in many cases was complicit in subverting women's daily security. As Enloe (2013: 80) states, 'it makes sexual harassment and domestic violence appear politically inconsequential'. Unfortunately, international pressure and attention to this issue from international actors – whether during Mubarak's era, that of the Muslim Brotherhood, or since Abdel Fattah el-Sisi's rule began – have been minimal. The Egyptian women's movement was severely divided under Mubarak and continued to be so in the post-Mubarak Muslim Brotherhood era. In fact, only five out of eight women remained in the new constitution-writing assembly in 2012 – four of whom were members of the Muslim Brotherhood (Al Agati and Senari 2012). Egyptian women have been forcibly pushed out of public spaces, have lost their pre-revolution electoral quota status, and *physically* continue to be used as a politicized weapon to instil fear and limit public life. Violence in post-conflict Egypt is on the rise, and a growing and frustrated youth bulge – composed predominantly of young men – is increasingly turning to conflict and crime.

Conclusion

Throughout this chapter we argue that a gendered analysis is essential to creating true safety in urban spaces for all residents of a city, regardless of their gender. We show that while the current literature on the confluence of peacebuilding and urban safety practice is nourishing a conversation about how peacebuilding and urban safety can learn from one another to improve their practice, it has largely done so without adequately incorporating gender. We also suggest that there is considerable space to bring gender into the discussion, which would improve the quality of analysis and be more in keeping with the principles underlying the discourse, as well as the operational effectiveness of policy and practice flowing from the confluence of urban safety and peacebuilding. Integration of a gender analysis into urban safety and peacebuilding programmes and initiatives is likely to improve safety for both women and men living in urban centres. It also creates space for a reinvigorated transformational approach to these fields that can contribute to overcoming oppressive patriarchal power structures and bring about more equal relations among men and women.

We argue that there may be five primary entry points for incorporating gender into practice that occur at the nexus of urban safety and peacebuilding: women's urban safety and security, masculinities, gender-sensitive SSR, involving women in peacebuilding processes, and supporting women's movements to politicize gender issues. Focusing on gender inequalities as a root cause of continued or recurring violence/conflict has the potential to transform gender relations and promote sustainable peace. However, if gender continues to be deprioritized in urban safety and peacebuilding practice, urban violence, social inequalities, and exclusion will persist, if not rise, and armed conflict is likely to resume.

Ultimately, there may be two options to strengthen urban safety and peacebuilding in cities. Drawing on urban safety's situational focus, gender could be incorporated in a way that focuses on the individual, aims to empower women to make them safer, and starts to unpack the gendered power dynamics that keep them insecure. This would involve focusing only on making cities safe for women. Alternatively, recasting oppressive social and institutional structures could be built upon to bring about gender-*transformative* change that reconstitutes the expectations and norms that prevent women and men from being free from fear. Moser (2016) argues that this type of change rests on collective action and institutional partnerships. Lessons from gender-sensitive peacebuilding, including bringing women into decision-making and supporting women's political activism, echo these points and are likely to contribute to making urban spaces (public and private) safer and more inclusive for all and in more sustainable and sustained ways.

Notes

1 For an overview of key resources about the confluence of urban safety and peacebuilding see www.gpplatform.ch/peacebuilding-resources.

2 Our focus in this chapter on women and men, femininities and masculinities is not meant to exclude lesbian, gay, bisexual, transgender, and intersex (LGBTI) people. Indeed, we recognize their importance and unique safety needs, and we encourage further research on LGBTI issues at the confluence of urban safety and peacebuilding.

3 The literature on the global North largely supports this assertion. For example, the Geneva Declaration Secretariat (2011: 130) notes that only 8 per cent of homicide

victims of domestic and spousal conflicts in the European Union in 2006 were male. Moreover, female perpetrators of domestic violence are much more likely to use violence in self-defence and in the context of an existing abusive relationship (Swan *et al.* 2008). Very little evidence on female perpetrators of violence outside intimate partnerships or in the developing world exists, although there is a modest body of work on women in armed conflict.

4 Meaningful mainstreaming of gender is a strong principle; in practice it has been largely superficial, if not harmful, to the cause of women's rights (see, for example, Moser and Moser 2005).

5 The Arabic acronym for the phrase al-Dawla al-Islamiya al-Iraq al-Sham (Islamic State of Iraq and the Levant).

6 Interviews were conducted in Cairo, Jakarta, and Bali in May and June 2013 with leading members of women's movements in both countries, and include interviews with Indonesian feminist academics (such as Chusnul Mar'iyah, who is using feminist interpretations of the Quran to counter religious discourse against women's rights), civil society actors, and members of Indonesia's parliament.

7 As recounted by Chusnul Mar'iyah during an interview in 2013.

References

Al Agati, M. and N. Senari. 2012. 'Women and Equal Citizenship: Analysis of the New Constitution of Egypt', Arab Forum for Citizenship in Transition, available at http://webserver1.thebody.com/about-us/ (accessed 10 October 2018).

Anderlini, S. N. 2007. *Women Building Peace: What They Do, Why It Matters.* Boulder, CO: Lynne Rienner.

Anwar, N. H., D. Mustafa, A. Sawas, and S. Malik. 2016. *Gender and Violence in Urban Pakistan.* Karachi and London: Institute for Business Administration and King's College London.

Arkles, G. and P. Gehi. 2016. 'Dreaming, Telling, Occupying, and Destroying: Interest Convergence between Militarism and Social Justice in the Dream Act and the Repeal of "Don't Ask, Don't Tell" ', *Scholar and Feminist Online* 13(2): np.

Assiago, J. 2016. Remarks at Habitat III Side Event, 'Creating Safe and Inclusive Cities that Leave No One Behind', Habitat III, 19 October, Quito, Ecuador.

Baranyi, S. and K. Powell. 2005. *Fragile States, Gender Equality and Aid Effectiveness: A Review of Donor Perspectives.* Ottawa, ON: North–South Institute, available at www.nsi-ins.ca/wp-content/uploads/2012/10/2005-Fragile-States-Gender-Equality-and-Aid-Effectiveness-A-Review-of-Donor-Perspectives.pdf (accessed 20 June 2018).

Barolsky, V., B. Roberts, S. Gordon, I. Cano, E. Ribeiro, D. Borges, P. Nunes, T. Duarte, and D. McLennan. 2016. 'Research Brief: Is Social Cohesion the Missing Link in Preventing Violence?' Pretoria: Human Sciences Research Council.

Bastick, M. and K. Valasek (eds). 2008. *Gender and Security Sector Reform Toolkit.* Geneva: DCAF, OSCE/ODIHR, and UN-INSTRAW.

Buss, D. E. 1997. 'Going Global: Feminist Theory, International Law, and the Public/Private Divide', in S. B. Boyd (ed.) *Challenging the Public/Private Divide: Feminism, Law, and Public Policy.* Toronto, ON: University of Toronto Press, 360–384.

Chetail, V. and O. Jütersonke. 2015. 'Peacebuilding: A Review of the Academic Literature', White Paper Series 13. Geneva: Geneva Peacebuilding Platform.

Connell, R. W. and J. W. Messerschmidt. 2005. 'Hegemonic Masculinity: Rethinking the Concept', *Gender and Society* 19(6), 829–859.

Dymén, C. and V. Ceccato. 2012. 'An International Perspective of the Gender Dimension in Planning for Urban Safety', in V. Ceccato (ed.) *The Urban Fabric of Crime and Fear*. New York: Springer, 311–339.

El-Bushra, J. 2012. *Gender in Peacebuilding – Taking Stock.* London: International Alert.

Enloe, C. 1983. *Does Khaki Become You?* London: Pluto.

Enloe, C. 1989. *Bananas, Beaches and Bases: Making Feminist Sense of International Politics.* London: Pandora Press.

Enloe, C. 1993. *The Morning After: Sexual Politics at the End of the Cold War*. Oakland, CA: University of California Press.

Enloe, C. 2013. 'Masculinities, Policing, Women and International Politics of Sexual Harassment', *International Feminist Journal of Politics* 15(1), 77–81.

Eriksson Baaz, M. and M. Utas. 2012. 'Beyond "Gender and Stir": Reflections on Gender and SSR in the Aftermath of African Conflicts', Policy Dialogue No. 9. Uppsala: Nordic Africa Institute.

Galtung, J. 1964. 'An Editorial', *Journal of Peace Research* 1(1), 1–4.

Geneva Declaration Secretariat. 2011. *Global Burden of Armed Violence 2011: Lethal Encounters*. Geneva: Geneva Declaration Secretariat.

Geneva Declaration Secretariat. 2015. *Global Burden of Armed Violence 2015: Every Body Counts*. Geneva: Geneva Declaration Secretariat.

Goetz, A. M. and R. Jenkins. 2015. 'Agency and Accountability: Promoting Women's Participation in Peacebuilding', *Feminist Economics* 22(1), 211–236.

Grob, A., A. Papadovassilakis, and L. Ribeiro Fadon Vicente. 2016a. 'Entry Points for Programming across Urban Safety and Peacebuilding', Paper 16. Geneva: Geneva Peacebuilding Platform.

Grob, A., A. Papadovassilakis, and L. Ribeiro Fadon Vicente. 2016b. 'Case Studies on Urban Safety and Peacebuilding: Lagos, Beirut, Mitrovica, Treichville and Johannesburg', Paper 19. Geneva: Geneva Peacebuilding Platform.

Hale, S. 2001. 'The State of the Women's Movement in Eritrea', *Northeast African Studies* 8(3), 155–177.

Handrahan, L. 2004. 'Conflict, Gender, Ethnicity, and Post-Conflict Reconstruction', *Security Dialogue* 35(4), 429–445.

Hans, A. and S. Rajagopalan (eds). 2016. *Openings for Peace: UNSCR 1325, Women and Security in India.* Thousand Oaks, CA: Sage Publications.

Harders, C. 2011. *Gender Relations, Violence and Conflict Transformation. Advancing Conflict Transformation: The Berghof Handbook II.* Opladen and Farmington Hills, MI: Barbara Budrich Publishers.

Harland, K. and S. McCready. 2014. 'Rough Justice: Considerations on the Role of Violence, Masculinity, and the Alienation of Young Men in Communities and Peacebuilding Processes in Northern Ireland', *Youth Justice* 14(3), 269–283.

Humansecurity-cities.org. 2006. 'Human Security for an Urban Century: Local Challenges, Global Perspectives', available at https://docs.unocha.org/sites/dms/HSU/human_security_for_an_urban_century%20South%20America%20%281%29.pdf (accessed 20 June 2018).

ICPC (International Centre for Prevention of Crime). 2016. 'Chapter 2: Urban Safety', in *ICPC Fifth International Report. Crime Prevention and Community Safety: Cities and the New Urban Agenda.* Montreal, QC: ICPC.

IDRC (International Development Research Centre). 2017. 'Gender and Violence in Cities', Ottawa, ON: IDRC, available at www.idrc.ca/en/research-in-action/gender-and-violence-cities (accessed 24 June 2018).

Jouejati, R. 2017. 'Women Are Invisible at the Syria Peace Talks', *PassBlue*, available at www.passblue.com/2017/02/23/whos-missing-at-the-syria-peace-talks-women/ (accessed 24 June 2018).

Jütersonke, O. and K. Krause. 2013. 'Peacebuilding and the City: Setting the Scene', Brief 9. Geneva: Geneva Peacebuilding Platform.

Kabeer, N. 1994. *Reversed Realities: Gender Hierarchies in Development Thought.* London: Verso.

Klot, J. F. 2007. 'Women and Peacebuilding', independent expert paper commissioned by UN Development Fund for Women (UNIFEM) and Peacebuilding Support Office. New York: United Nations.

Lock, K. 2016. 'Women in the Syrian Peace Process: More Than a Seat at the Table?', UNU-MERIT blog. Maastricht: UNU, available at www.merit.unu.edu/women-in-the-syrian-peace-process-more-than-a-seat-at-the-table/ (accessed 20 June 2018).

Mahadevia, D. and R. Desai. Forthcoming 2019. 'Everyday Violence in India: Is Planing the Driver or Mitigator?', in J. E. Salahub, M. Gottsbacher, J. de Boer, and M. D. Zaaroura (eds) *Reducing Urban Violence in the Global South: Towards Safe and Inclusive Cities*. London: Routledge.

Mahadevia, D., R. Desai, S. Sanghvi, S. Vyas, and V. Parmar. 2017. 'Vatwa Resettlement Sites: Gender Insecurity and Violence against Women', Safe and Inclusive Cities, Poverty, Inequality, and Violence in Indian Cities: Towards Inclusive Policies and Planning Ahmedabad, Policy Brief 5. Ahmedabad: CEPT University.

Mahadevia, D., R. Desai, S. Sanghvi, S. Vyas, A. Bakde, M. S. Pathan, and R. Malek. 2016. 'Bombay Hotel: Gender Insecurity and Violence against Women', Safe and Inclusive Cities, Poverty, Inequality, and Violence in Indian Cities: Towards Inclusive Policies and Planning Ahmedabad, Policy Brief 9. Ahmedabad: CEPT University.

Meintjes, S., M. Turshen, and A. Pillay. 2001. *The Aftermath: Women in Post-Conflict Transformation.* London: Zed Books.

Milliken, J. 2013. 'Building Peace in Urban Setting: Lessons from the Ground and Recommendations for Collaboration, Report of the 2013 Annual Meeting', Paper 11. Geneva: Geneva Peacebuilding Platform.

Milliken, J. 2016. 'Urban Safety and Security: Lessons from the Last Two Decades and Emergent Issues', Paper 22. Geneva: Geneva Peacebuilding Platform.

Moser, C. O. N. 2004. 'Urban Violence and Insecurity: An Introductory Roadmap', *Environment and Urbanization* 16(2), 3–16.

Moser, C. O. N. 2016. 'Gender Transformation in a New Global Urban Agenda: Challenges for Habitat III and Beyond', *Environment and Urbanization* 29(1), 221–236.

Moser, C. O. N. and F. Clark. 2001. *Victims, Perpetrators, or Actors? Gender, Armed Conflict, and Political Violence.* London: Zed Books.

Moser, C. O. N. and A. Moser. 2005. 'Gender Mainstreaming since Beijing: A Review of Success and Limitations in International Institutions', *Gender and Development* 13(2), 11–22.

Muggah, R. 2012. *Researching the Urban Dilemma: Urbanization, Poverty, and Violence*. Ottawa, ON: IDRC.

Munro, J. 2000. *Gender and Peacebuilding*. Ottawa, ON: IDRC.

Myrttinen, H., J. Naujoks, and J. El-Bushra. 2014. *Re-thinking Gender in Peacebuilding.* London: International Alert.

Nilsson, D. 2012. 'Anchoring the Peace: Civil Society Actors in Peace Accords and Durable Peace', *International Interactions* 38, 243–266.

O'Leary, R. and K. Viswanath. 2011. *Building Safe and Inclusive Cities for Women: A Practical Guide*. New Delhi: Jagori.

Owusu, G. 2016. 'Introduction: Urban Crime and Poverty Nexus', *Ghana Journal of Geography* 8(1), 1–10.

Paffenholz, T., N. Ross, S. Dixon, A. Schluchter, and J. True. 2016. *Making Women Count – Not Just Counting Women: Assessing Women's Inclusion and Influence on Peace Negotiations*. Geneva: Inclusive Peace and Transition Initiative and UN Women.

Pankhurst, D. 2010. 'The 'Sex War' and Other Wars: Towards a Feminist Approach to Peacebuilding.' *Development in Practice* 13(2–3): 154–177.

Pankhurst, D. 2016. '"What is Wrong with Men?" Revisiting Violence Against Women in Conflict and Peacebuilding.' *Peacebuilding* 4(2): 180–193.

Parent, A. 2016. 'The Programme Conjoint de Renforcement de la Sécurité Urbaine in Ouagadougou, Burkina Faso: Lessons Learned', Paper 20. Geneva: Geneva Peacebuilding Platform.

Rathgeber, E. M. 1990. 'WID, WAD, GAD: Trends in Research and Practice', *Journal of Developing Areas* 24(4), 489–502.

Rehn, E. and E. J. Sirleaf. 2002. *Women, War and Peace: The Independent Experts' Assessment on the Impact of Armed Conflict on Women and Women's Role in Peace-Building*. New York: UNIFEM.

Rosenthal, G. (ed.). 2015. *The Challenge of Sustaining Peace: Report of the Advisory Group of Experts for the 2015 Review of the United Nations Peacebuilding Architecture*. New York: United Nations.

Saferworld. 2014. *Masculinities, Conflict and Peacebuilding: Perspectives on Men Through a Gender Lens*. London: Saferworld.

Sáiz, E. and B. Lasagabaster. 2017. 'We Need Global Data on Women's Participation in Local Government', Citiscope commentary, available at http://archive.citiscope.org/commentary/2017/03/we-need-global-data-womens-participation-local-government (accessed 23 June 2018).

Salahub, J. E. and K. Nerland. 2010. 'Just Add Gender? Challenges to Meaningful Integration of Gender in SSR Policy and Practice', in M. Sedra (ed.) *The Future of Security Sector Reform*. Waterloo, ON: Centre for International Governance Innovation, 261–278.

Salazar Sanchez, K. and J. P. Pérez Sáinz. 2015. 'Estrategias de Confrontación ante Algunas Formas de Violencia: La Experiencia de Dos Territorios Costarricenses', in J. P. Pérez Sáinz (ed.) *Exclusión Social y Violencias en Territorios Urbanos Centro americanos*. San José: FLACSO.

Shakir, W. 2015. *Women and Peacemaking in Yemen: Mapping the Realities*. Helsinki: Crisis Management Initiative.

Slegh, H., E. Mariano, S. Roque, and G. Barker. 2016. *Being a Man in Maputo: Masculinities, Economic Stress and Urban Violence in Maputo, Mozambique*. Rio de Janeiro and Washington, DC: Instituto Promundo.

Small Arms Survey. 2012. 'Women and Armed Violence in South Sudan', Sudan Human Security Baseline Assessment. Geneva: Small Arms Survey, available at www.smallarmssurveysudan.org/fileadmin/docs/facts-figures/south-sudan/womens-security/HSBA-women-and-armed-conflict.pdf (accessed 20 June 2018).

Sudhakar, N. and K. Kuehnast. 2011. 'The Other Side of Gender: Including Masculinity Concerns in Conflict and Peacebuilding', USIP PeaceBrief 75. Washington, DC: US Institute of Peace.

Swan, S. C., L. J. Gambone, J. E. Caldwell, T. P. Sullivan, and D. L. Snow. 2008. 'A Review of Research on Women's Use of Violence with Male Intimate Partners', *Violence Victims* 23(3), 301–314.

Sweetman, C. 2005. *Gender, Peacebuilding, and Reconstruction*. Oxford: Oxfam GB.

Taylor, A., T. Moura, J. Scabio, E. Borde, J. Afonso, and G. Barker. 2016. *'This Isn't the Life for You': Masculinities and Nonviolence in Rio de Janeiro, Brazil*. Rio de Janeiro and Washington, DC: Instituto Promundo.

Turshen, M. 1998. 'Women's War Stories', in M. Turshen and C. Twagiramariya (eds), *What Women Do in Wartime: Gender and Conflict in Africa*. London: Zed Books, 1–26.

UN-Habitat (UN Human Settlements Programme). 2007. *Enhancing Urban Safety and Security: Global Report on Human Settlements 2007*. London: Earthscan.

UN-Habitat (UN Human Settlements Programme). 2016. 'Safer Cities Programme', available at http://unhabitat.org/urban-initiatives/initiatives-programmes/safer-cities/ (accessed 20 June 2018).

UNSC (UN Security Council). 2000. Resolution 1325, 'Women, Peace and Security', UN Doc. S/RES/1325, 31 October. New York: United Nations.

UNSC (UN Security Council). 2013. Resolution 2122, 'Strengthening Women's Role', UN Doc. S/RES/2212, 18 October. New York: United Nations.

UNSC (UN Security Council). 2015. Resolution 2242, 'Improving Implementation of 1325', UN Doc. S/RES/2242, 13 October. New York: United Nations.

UNSC (UN Security Council). 2016. Resolution 2282, 'Post-Conflict Peacebuilding', UN Doc. S/RES/2282, 27 April. New York: United Nations.

UNSG (UN Secretary-General). 1992. 'An Agenda for Peace', UN Doc. A/47/277, 17 June. New York: United Nations.

UN Women. 2016a. 'Facts and Figures: Peace and Security', available at www.unwomen.org/en/what-we-do/peace-and-security/facts-and-figures (accessed 20 June 2018).

UN Women. 2016b. 'Facts and Figures: Ending Violence against Women', available at www.unwomen.org/en/what-we-do/ending-violence-against-women/facts-and-figures (accessed 20 June 2018).

UN Women. Nd. *In Brief: Safe Cities Global Initiative*. New York: United Nations.

van den Eertwegh, H. 2016. 'Negotiating with Criminal Groups: From Prejudice to Pragmatism', Paper 18. Geneva: Geneva Peacebuilding Platform.

Väyrynen, T. 2010. 'Gender and Peacebuilding', in O. P. Richmond (ed.) *Palgrave Advances in Peacebuilding: Critical Developments and Approaches*. New York: Palgrave Macmillan, 137–153.

Wennmann, A. 2016a. 'Building Peace: An Annotated Resource Pack', Paper 21. Geneva: Geneva Peacebuilding Platform.

Wennmann, A. 2016b. 'Sustaining Peace in the City: Perspectives on Urban Safety', conference brief. Geneva: Geneva Peacebuilding Platform.

Wennmann, A. and B. Ganson. 2016. 'Responses to Conflict that Work', Paper 15. Geneva: Geneva Peacebuilding Platform.

WHO (World Health Organization). 2014. *Global Status Report on Violence Prevention 2014*. Geneva: World Health Organization.

WICI (Women in Cities International). 2012. *Tackling Gender Exclusion: Experiences from the Gender-Inclusive Cities Programme*. Montreal, QC: WICI.

7 The intersection of community crime prevention and security sector reform

The case of Jenin

Peter Homel and Nicolas Masson

Introduction

In situations where central authorities are weakened by a lack of political legitimacy, systemic dysfunctions, corruption, or foreign interference, donor-supported institutional reforms often fail to trigger positive change for communities and citizens. This is true for security sector reform (SSR) programmes that promote state-centred, top-down reform of the armed and security forces by focusing on technical rather than political transformation. This chapter highlights some of the shortcomings of these top-down SSR programmes, in particular those arising from a neglect of citizens' views and needs, a lack of local ownership of the reform process, a failure to address broader governance issues, and the impact of a cooptation of political and security elites.

This chapter also describes possible avenues for overcoming these shortcomings by presenting a form of SSR undertaken between 2008 and 2014 in the Governorate of Jenin in the north of the Occupied Palestinian Territories (OPT) on the West Bank. The Jenin project aimed at producing a more sustainable SSR approach by integrating principles and methods of outcome-based community safety and participatory local crime prevention planning processes. In contrast to state-centred SSR experiments, the Jenin community safety project is a bottom-up, community-based approach that builds effective 'partnerships' for crime prevention with both formal security providers (e.g. security forces, executive authorities, parliament arians, and governors' offices) and informal security providers (e.g. civil society, media, tribal and business leaders) to produce a workable mechanism by which a safer community with stronger local leadership might be promoted.

The chapter summarizes the intersections between these newer local community safety initiatives and SSR programmes in fragile and conflict-affected environments by exploring the practical implications to be drawn from the Jenin case study. There, a top-down, donor-led SSR programme was conducted by several donors between 2008 and 2014. During the same period, the authors of this article were involved in the development of a bottom-up community safety programme sponsored by the Geneva Centre for the Democratic Control of Armed Forces (DCAF) and the local human rights organization SHAMS (see also Homel and Masson 2016).

To elaborate on the intersection between community crime prevention and SSR, the chapter first describes the changing donor contexts and clarifies key concepts and how they interrelate, before providing a detailed description of the lessons from the application of bottom-up community safety programmes in Jenin. The final section explores the overarching lessons and observations that may provide inspiration and guidance for other initiatives.

Conceptual clarifications

The global policy context

The international community has increasingly recognized the close link between effective crime prevention and fair, transparent, and humane criminal justice systems as key elements of achieving sustainable development, promoting the rule of law, and pursuing a state-building agenda (UNGA 2014). This means that security and justice are viewed as both enablers and outcomes of sustainable socioeconomic development and the creation of viable states. Societies that enjoy socioeconomic equality, good governance, and solid rule of law often demonstrate lower levels of crime, violence, and victimization. On the other hand, where societies are characterized by high inequality, violence, unemployment, and weak social fabric, states often fail in their obligations to respect, protect, and fulfil human rights, potentially feeding into a circle of more insecurity and development stagnation, or even losses in development gains (Jackson 2011; UNODC 2010).

This has led international agencies increasingly to support the use of what are in effect local community safety initiatives to promote human security and peacebuilding within unstable contexts (Baker 2010a; Mac Ginty 2015). Examples include activities such as decentralized multistakeholder public consultations on security in Tunisia (Harzallah and Masson 2014), the promotion of local peace agreements between municipal actors in Libya (PCI 2013), the establishment of local security councils in Kosovo, Nepal, and Bangladesh (Saferworld 2013), and even mediation initiatives between criminal gangs and local governance bodies as witnessed in Latin and Central America (Cocayne 2013; see also Chapter 4 in this volume).

Development agencies particularly focus on promoting local community safety in countries where state-centred security structures are ineffective. Where the authority of the state and its control over the security forces are weak or contested, it is likely that local communities will suffer from a security vacuum, potentially leading to abuses by informal power structures, including warlords or gangs, as has been noted particularly in places like Afghanistan (Sedra 2006) among others (Jackson 2011). Inclusive local community safety programmes aim at providing opportunities for security and justice to be fairly administered by local entities in the absence of a solid state framework (Gordon 2014).

These developments are a response to an emerging concern about the ability of the various forms of SSR models used in different contexts to address adequately the complex and apparently at times competing tasks of state building and ensuring the specific security needs of local communities (Hänggi and Scherrer 2008; Sedra 2010; Donais 2009). However, as Mac Ginty (2015) points out, simply talking in terms of strengthening the focus on 'localism' is not in itself a complete or adequate solution to this tension, as the roots of the problems that the SSR process seeks to address can often be traced to local communities themselves. This is one of the reasons why peacebuilding actors are starting to turn to the tools and processes developed for use in the crime prevention sector for the specific purpose of addressing local community conflict and to enhance local community safety.

Community safety and crime prevention processes

For the United Nations (UN), the notion of 'community safety' or 'community security' initially implied the protection of threatened ethnic minorities and indigenous communities, understood as groups sharing the same identity markers and values (UNDP 1994). More recently, community safety has increasingly been associated with the 'freedom from fear' or 'free from want' definition of human security (ibid.: 22, 24; Redo 2012). Community safety programmes are people-centred approaches to security that include strengthening local communities' capacity to respond to armed violence and crime ('freedom from fear'). Local communities are empowered to build stronger partnerships with state authorities and security institutions with a view to tackling a wide range of political, socioeconomic, environmental, and health security issues ('free from want'). Community safety is also consistent with the UN promotion of the principle of 'subsidiarity', according to which 'issues ought to be addressed at the lowest level capable of addressing them', i.e. at the local, national, subregional, and regional levels, while 'reducing the number of issues that need to be tackled at the international and supranational level' (United Nations 2014: 14).

The process of crime prevention is a fundamental tool for the achievement of community safety. The UN *Guidelines for the Prevention of Urban Crime* define crime prevention as comprising those 'strategies and measures that seek to reduce the risk of crimes occurring, and their potential harmful effects on individuals and society, including fear of crime, by intervening to influence their multiple causes' (ECOSOC 2002). What is important about this definition is that crime prevention is clearly differentiated from action intended to simply control crime. Today, international experience has shown that effective crime prevention can maintain and reinforce the social cohesion of communities and assist them to act collectively to improve their quality of life (Idriss *et al.* 2010; Sampson 2013; Cherney 2017).

In recent years a rigorous approach to local community crime prevention has emerged, frequently built around the use of highly local governance structures such as local government authorities (Homel and Fuller 2015; Morgan *et al.* 2012; Morgan and Homel 2011; Cherney 2017). These new local community crime prevention approaches also place a stronger emphasis on more rigorous strategic planning processes built on ensuring high levels of local governance and community participation as part of a goal to ensure transparency and strong accountability through the use of performance measures and outcome-based planning (Morgan and Homel 2011).

As an overall approach, crime prevention comprises a range of strategies implemented by individuals, communities, businesses, non-governmental organizations (NGOs), and all levels of government to target the various individual, social, and environmental factors that increase the risk of crime, disorder, and victimization (AIC 2003; ECOSOC 2002; IPC 2008; van Dijk and de Waard 1991). This includes strategies that modify the physical environment to reduce the opportunities for crime to occur (environmental approaches), and strategies that focus on the underlying social and economic causes of crime and limiting the supply of motivated offenders (social approaches; see also Chapters 1 and 2 in this volume). Crime prevention may have a universal focus on strengthening institutions that support civil society and the rule of law, be directed at the early identification and subsequent intervention in the lives of people or groups at risk of engaging in criminal activity, or be targeted at the prevention of recidivism among those who have already engaged in offending behaviour (Welsh and Farrington 2012).

There are a range of actions delivered in other sectors (e.g. health, education, housing, and human services) that will have an impact on crime levels, especially in the long term, but may not have the prevention of crime as their primary goal. While these activities do not necessarily fall within a strict definition of crime prevention, it is important that the potential crime prevention benefits of the policies and programmes delivered in these sectors are identified, acknowledged, and, wherever possible, enhanced.

Security sector reform

SSR generally refers to a process to reform or rebuild a state's security sector. It responds to a situation in which a dysfunctional security sector is unable to provide security to the state and its people effectively and under democratic principles. In some cases the security sector can itself be a source of widespread insecurity due to discriminatory and abusive policies or practices. In this respect, an unreformed or poorly constructed security sector represents a decisive obstacle to the promotion of sustainable development, security, and peace. SSR processes therefore seek to enhance the delivery of effective and efficient security and justice services by security sector institutions that are accountable to the state and its people, and that operate within a framework of democratic governance, without discrimination, and with full respect for human rights and the rule of law (Ball 2010).

SSR is both an operational and a normative concept. Featuring norms such as good governance, civilian oversight, and the rule of law among its defining characteristics, its inclusion as a necessary component of international policies addressing post-conflict situations is becoming more and more commonplace.

As such, SSR can be seen as a branch of an increasing international effort to promote human security (Bryden and Hänggi 2005).

The UN defines SSR as a transformative process aiming to create 'effective and efficient state security forces, capable of providing security for the state and its people, within a framework of democratic civilian control, rule of law, and respect for human rights' (UNDPKO 2012: 2). Research in development affairs establishes a clear link between development and security, defined as a basic right alongside other services such as health or education. Comprehensive SSR programmes include reforming democratic oversight institutions such as parliament, the judiciary, and the media. This broader approach to SSR supports 'sustainable development and poverty reduction' by making security providers more efficient and accountable to the people (OECD 2008: 15); but it stands in contrast to the narrower 'train–build–equip' vision of SSR supported by some donors that does not encompass democratic governance and oversight of the security forces (Ball 2010; Hänggi and Scherrer 2008).

A case study of SSR in Jenin Governorate, OPT (2008–2014)

Jenin is the northernmost of the 11 Palestinian governorates in the OPT. It has a population of 260,000 and borders to the north and west with Israel and to the east with Jordan. Since the 1993 Oslo peace agreements the governorate and the remainder of the West Bank have been divided into three administrative and security zones. In Zone A Palestinian authorities have full control over services, including security; in Zone B this control is shared between Palestinian authorities and the Israel Defence Force (IDF); and in Zone C the IDF exerts full security control. Zone C areas cover 60 per cent of the West Bank and are mostly adjacent to the borders with Israel and Israeli settlements; 2,300 Israeli settlers live in five main settlements within the boundaries of Jenin Governorate.

During the Second Intifada (2000–2005) Jenin was at the heart of some of the fiercest battles between the IDF and armed Palestinian factions. After the end of the Second Intifada and following the takeover of the Gaza Strip by Hamas-affiliated brigades in June 2007, major international donors (the USA, the European Union (EU), Russia, and the UN, known as the 'Quartet') started a security campaign aimed at reorganizing the scattered Palestinian security apparatus. Initiated in early 2008 in Jenin, this campaign was called Operation Hope and Smile. It involved thousands of US-trained Palestinian security officers, and had the objective of restoring law and order in the governorate. Local and international observers have underlined some of the successes of this campaign. The Israeli and US press praised the 'Jenin model' as an opportunity to foster 'a bottom-up set of relationships between the Palestinian and the Israeli societies' (Bronner 2008). As it resulted in the arrest of 'scores of Hamas members and suspected sympathizers,' other observers criticized the campaign for pursuing politically biased 'anti-terror' objectives (ICG 2010: 6).

As underlined by Friedrich and Luethold (2009), the international donors misleadingly labelled the security campaigns conducted in the West Bank in

2008–2009, such as 'Hope and Smile', as 'SSR'. As the key shortcomings highlighted here show, these campaigns were inconsistent with the internationally adopted SSR standards of the Organisation for Economic Co-operation and Development (OECD). These standards imply 'the establishment of effective governance, oversight and accountability, so that the security forces and the political authorities, which control and oversee them, operate in a manner consistent with democratic norms, and within the rule of law' (ibid.: 199).

The so-called 'SSR' initiatives in Jenin and the rest of the West Bank occurred at the same time as development experts and practitioners were increasingly questioning the sustainability of state-centred SSR programmes in fragile contexts. As observed in places such as Lebanon (Larzillière 2012: 15), Kosovo, and Timor-Leste (Gordon 2014: 2–3), donor-led, top-down SSR processes are often characterized by poor integration of the security needs and expectations of citizens, limited inclusiveness of the strategic decision-making process, and a lack of locally entrenched governance and democratic oversight frameworks.

Research on the security campaigns in the Palestinian West Bank also tends to highlight the shortcomings of donor-driven, top-down SSR processes in unstable contexts. The key shortcomings identified by this research with respect to Jenin emphasizes, for instance, the absence of a shared strategic national security objective. The donors' support of a non-inclusive SSR programme for the Palestinian Authority's security and political apparatus under military occupation, without a credible long-term strategic objective, has left Palestinian security forces with the unsustainable contradiction of working 'with the occupier in providing security for the occupied' (ICG 2010: 40). What is more, there was a lack of local ownership of the reform process. The externally led Palestinian SSR programme focused on training, building, and equipping Palestinian security forces rather than developing a governance framework for the Palestinian security sector. Donors led a force-to-force technical-tactical 'transformation of the security apparatus' involving their military, police, and intelligence officials rather than fostering a local reform process which the Palestinians could 'own' with a view to strengthening their future state structures (Friedrich and Luethold 2009: 199).

The research also underlines the failure to address key governance issues as a shortcoming of the security campaigns. The blurring and overlapping of mandates between the military-like US-trained Palestinian National Security Forces and the EU-supported Palestinian Civil Police remained a key problem throughout the SSR programme. As a Palestinian security officer from Jenin observed in 2009, 'the police only should have the right to arrest people. It is inappropriate to have so many security forces carrying out arrests' (DCAF 2009: 3). Such efforts have implicitly fostered a more authoritarian rather than democratic transformation of the security sector. The donors' support for an 'authoritarian transformation' encouraged the development of a security regime in the occupied West Bank rather than promoting a democratic transition based on the transparency and accountability of the security sector (Seyigh 2011: 19). In consequence, transformation was driven more by a process of cooptation than by local empowerment. The SSR programme in Jenin furthered the powerful international players' cooptation of local elites and Palestinian political fragmentation instead of institutionalized rules, permanent laws, and a genuine vision of security as a public good (Marten 2014: 181).

Lessons learnt from bottom-up community safety processes in Jenin

In many developed countries strategic models for community safety and crime prevention programmes are increasingly adopted to address insecurity, mostly in urban centres (Idriss *et al.* 2010). Since the 1980s, long before international aid agencies started promoting local community safety programmes in fragile contexts, criminologists had questioned the efficiency of 'hard' top-down responses to address citizens' insecurity (Welsh and Farrington 2012). While investigating, prosecuting, and punishing crime remained important, issues pertaining to health, education, and socioeconomic factors became increasingly part of community-based crime prevention programmes. Criminologists initiated a widening and deepening of the approaches to crime prevention, promoting community participation and local partnerships for tackling insecurity (Homel 2005).

In 2011 the Australian Institute of Criminology (AIC) published a model performance framework for community-based crime prevention processes (Morgan and Homel 2011) designed to support Western Australian local authorities in their crime prevention planning. This model offers an overview of how participative strategic community-based crime prevention processes can operate (ibid.: 28). While essentially a problem-solving approach, the AIC model places a strong emphasis on the establishment of clearly defined and agreed-upon goals. This is something that is generally quite challenging in communities featuring strong and competing interest groups with local access to different levels and forms of power, regardless of their levels of cohesion and existing capacity. However, experience has shown that in deeply divided communities with access to few resources and little or no crime prevention experience or capacity, the approach provides a forum for galvanizing the development of at least a shared, if not always common, purpose. In addition the model places an emphasis on ensuring that once these goals are agreed, they are linked to well-articulated outcomes for which performance measures are also identified and embedded in the early stages of the planning process. It is made clear that these indicators need not all be purely quantitative measures, as these are frequently either unavailable or simply not accessible for various reasons (Homel and Fuller 2015; Hulme and Homel 2015). Invaluable for accountability purposes, measurement also helps to promote a transparent process where all stakeholders have access to information about programme performance during implementation.

As such, the Morgan and Homel (2011) model closely reflects the principles of good governance for partnership operation widely employed in the international development sector (Edgar *et al.* 2006), and later adapted for use in the development and management of crime prevention partnerships by Homel and Homel (2012). There are five elements to these principles for effective partnerships.

- Legitimacy and voice. A need to ensure that those in power are perceived to have acquired their power legitimately, and there is an appropriate voice accorded to those whose interests are affected by decisions.
- Direction and strategic vision. The exercise of power results in a sense of overall direction that serves as a guide to action.
- Performance. Institutions and processes are responsive to the interests of participants, citizens, and stakeholders.
- Accountability. There is accountability between those in positions of power and those whose interests they serve, and transparency and openness in the conduct of the work.
- Fairness. There is conformity with the rule of law and the principle of equity.

It was the capacity of this approach to community-based crime prevention to provide alternatives to 'hard' security measures against crime and insecurity that led to the decision to employ the Morgan and Homel (2011) model as the working basis for designing and implementing the DCAF and SHAMS community safety project in Jenin.

The four phases of the community safety planning processes described in the model (partnership agreement; community safety committee formation; research and public consultation; and developing a plan) proved relevant in Jenin, although in a different sequence. In Jenin the process started with public consultations, which led to the formation of a community safety committee whose members agreed on a partnership framework and eventually participated in developing a community safety plan (DCAF 2012: 8). This consultative phase of the community safety programme entailed a multistakeholder analysis of local security needs in Jenin that was shaped by different types of exchanges.

- Town hall meetings gathering over 100 Palestinian stakeholders from civil society, executive, legislative, and judicial authorities, and security officers under the title 'Delivering Security to the Palestinian People' (DCAF 2009).
- Focus group discussions with representatives of the Palestinian security forces, civil society, and tribal leaders under the title 'Strengthening the Rule of Law in the Governorate of Jenin' (DCAF 2010a).
- Focus group discussions with female community representatives to assess the gender-specific security needs of Palestinian women and girls (DCAF 2010b).
- Bilateral discussions with key community leaders and representatives of the authorities.

The broad consultations with local stakeholders confirmed the gap between the security needs of the Palestinian citizens and the strategic objectives of the top-down SSR programme conducted by the international community in Jenin. Key inconsistencies between the community-level expectations and the objectives of the security campaigns piloted by US- and EU-trained Palestinian security forces appeared at several levels.

The first inconsistency played out with respect to the role of local community engagement in the SSR process. There were high levels of ambiguity and a lack of understanding among the citizens in Jenin about the international SSR programme. As one of them asked, 'are the security forces here to protect Israel or to protect the Palestinian citizens and to establish a state?' (DCAF 2009). Because they were not formally consulted, citizens in Jenin did not see the high-stakes political dividend they could draw from cooperating with the donor-sponsored Palestinian security forces in establishing security in Jenin. This gap between the people's security needs and expectations on the one hand and the security agenda of the international community on the other generated mutual mistrust, undermined the reputation of the Palestinian Authority, eroded the credibility of its security forces, and remained the most fundamental obstacle to the SSR programme's sustainability. The community engagement process that was integral to the local crime prevention planning process worked to overcome this strategic deficit.

The second inconsistency was about institutional capacity. The donor-driven SSR programme failed to address key governance issues. For instance, there was no credible process for reviewing the legal framework of the Palestinian security forces. The security legislation remained a patchwork of old and new, non-democratic legislation and arbitrary presidential decrees (Friedrich *et al.* 2008). As a security officer from Jenin put it, 'we are in need of better laws in order to have a clear basis for our work' (DCAF 2009). Furthermore, it also appeared that donor-led parallel reforms of the Palestinian judiciary and empowerment of Palestinian civil society organizations were not integrated into the SSR process. Participants in the consultations complained that the Palestinian parliament (which was boycotted by most Quartet members in 2006 after Hamas won the legislative elections), the judiciary, and the local media did not exert serious oversight of the Palestinian security forces' operations. Defining governance aspects of the security of Palestinian citizens was not considered as part of the Palestinian security agenda. Again, it was through the crime prevention planning process that these issues could surface and be discussed in an environment that facilitated their inclusion in the SSR process.

The third inconsistency related to operational effectiveness. Operationally the participants expected that measures would be taken to curb human rights violations and abuses of power by the Palestinian security forces. They denounced the perpetration of political and extrajudicial arrests and detention by the security forces trained by the international community. The Jenin bottom-up community safety consultative process raised the question of how and to what extent local security needs and expectations can be brought to intersect with top-down SSR processes such as the one described above. The experience in Jenin thus brought a focus on the relevance of the two notions of 'ownership' and 'partnership' for creating such intersection opportunities between community-based crime prevention and top-down SSR processes. This is discussed in the next section.

Partnerships rather than local ownership?

Enshrined in key development charters such as the 2005 Paris Declaration on Aid Effectiveness (OECD 2008: 3) and the OECD Development Assistance Committee's *Handbook on Security System Reform* (OECD 2007: 64), the principle of local ownership implies that developing countries lead their own development policies and strategies, and integrate them in their own governance mechanisms (for instance, their national budget documents). In relation to SSR specifically, local ownership means in practical terms 'that the reform of security policies, institutions and activities in a given country must be designed, managed and implemented by local actors rather than external actors' (Nathan 2006: 8).

The question of the local ownership of peace and security processes, however, is very controversial (Donais 2009; Mac Ginty 2015; Baker 2010a; Gordon 2014). Problematic aspects of the concept of local ownership in relation to SSR include the following.

- The difficulty inherent in the SSR process of defining 'what exactly should be "owned" by which "locals"' (Mannitz 2014: 274; Mobek 2010). For instance, the legitimacy of local security stakeholders is often difficult to determine (Richmond 2012).
- The observation that in-depth involvement with local communities in unstable contexts implies dealing with entities such as traditional justice systems that do not meet democractic, human rights, and rule-of-law standards (Kritz 2007; Baker 2010b).
- The fact that local actors sometimes do not agree on minimal requisites for coexistence and peace. The absence of local consensus about the long-term objectives of the SSR programmes gives little opportunity for these actors to own such complex processes fully.
- The fact that high-stakes foreign-funded development programming takes place in the donor/expert community and de facto establishes a power patron–client relationship between external and local needs and interests, which makes local ownership virtually unpractical (Reich 2006: 4; Byrd 2010).
- Finally, there is sometimes a lack of 'capacity within partner governments to assess, design and implement reform processes' (OECD 2007: 14; Mac Ginty 2010).

Following the recommendations of the AIC crime prevention and community safety model, in Jenin the notion of 'partnership' was redefined as a more rewarding and less controversial guiding principle for the local community-based security process. The establishment of a strategic community safety partnership in Jenin was formalized by the creation of a community safety steering committee whose 15 members received training in guiding principles of strategic planning, community safety, and the rule of law. These efforts had several results.

- Setting the focus on the interactions between formal and informal security providers. In other words, rather than focusing on established actors and their roles in the Jenin community, the notion of 'partnership' set the focus on the mutual interaction between often-contending formal and informal actors (the 'community safety partners') who all needed a space for dialogue and exchange of views on security, including NGOs, women's rights centres, the media, tribal elders, security officers, and the governor's office.
- Reinforcing the security functions of local actors. This included formalizing partnerships for community safety in Jenin, which confirmed the importance of conferring 'recognition of existing local knowledge, capacities and skills with regards to particular security functions' (Mannitz 2014: 279).
- Fostering public support for security measures. This meant that formal security providers, such as the police and the established court system in Jenin, acknowledged that partnering with informal local stakeholders such as civil society organizations, media institutions, and tribal leaders played a key role in enforcing law and order.
- Enabling public information sharing on security. Through coverage by the local media of the community safety partners in Jenin, the central-level authorities in Ramallah (Council of Ministers, Office of the President) were informed of the process and eventually endorsed it officially.
- Creating a space for working donor–beneficiary relations. The Jenin community safety partnerships allowed the open acknowledgement that local communities (as is the case in developed countries such as Australia) do need conceptual and financial assistance to contribute to complex endeavours such as SSR or crime prevention programmes.

All of this in turn leads to propositions for practical intersections between community safety and SSR approaches where exchange of information and practices can be mutually reinforcing for bottom-up and top-down security approaches.

The practical intersections between community safety and SSR approaches

Community safety initiatives do not only inform SSR processes on citizens' perceptions of insecurity. SSR processes can benefit from integrating individual partners, groups, and existing mechanisms that aim to tackle insecurity at the local level (Gordon 2014). Indeed, such an approach is recommended by the UN *Security Sector Reform Integrated Technical Guidance Notes* (UNSSR Inter-Agency Task Force 2012). For the outcomes achieved in community-based crime prevention to be integrated into comprehensive national SSR strategies, it is critical to identify areas in which community-based security and key SSR objectives intersect effectively. Three potential intersection areas have been identified by the work in Jenin.

Strengthening the capacities of key security stakeholders

In their functions and behaviour, key 'agents of change' serve as transmission belts between the local and the national levels. Formal and informal community leaders, representatives of civil society, tribal or religious leaders, and local business people are in charge of transmitting the security concerns of their communities to central authorities. At the receiving end of national security policies, they are also the ones who contribute to making these policies understood and accepted by their communities. With a view to integrating local community-based security needs into broader security policies, it is important to create enabling mechanisms for these community leaders to play their role as agents of security transformation. Similar to what was achieved in Jenin, the integration of local leaders into structured community safety processes is a first important step towards establishing an effective intersection with top-down SSR policies. This integration process can include the following.

- Mapping the key local actors and their role in the community in terms of security. In Jenin this included tribal leaders involved in traditional mechanisms of conflict resolution, but also women's rights advocates, refugee camp committee leaders, and village council mayors.
- Organizing consultation sessions in which these key stakeholders express their views about their communities' security expectations; summarizing and structuring the key discussion points and transmitting them to decision-makers; and providing feedback to the community on the decision-makers' opinion for ensuring broad consensus.
- Identifying structured channels for informing central authorities about the contents of these consultations. This can be done through either bilateral talks or advocacy activities (publications, media coverage) – as in Jenin, where one of the key partners was a media and advocacy organization (SHAMS).
- Training formal and informal security actors at the community level in standards of democratic SSR processes, including key principles of the rule of law, conflict sensitivity, and conflict resolution mechanisms.
- Involving community leaders in surveys about the security perceptions of their community, especially with regards to vulnerable groups such as women, refugees, and young unemployed people.

Developing regulatory frameworks for the security sector

In many unstable contexts, local customs often clash with the official constitutional and legal framework (Mac Ginty 2015). Reforming the legal framework for the provision of security, which is a key element of SSR, may be hampered by persisting with traditional ways of providing security and justice. In Jenin, for instance, many perceived tribal conflict resolution mechanisms are more efficient than the institutional justice and court system. Other key elements of SSR, like the establishment of an independent judiciary, the conduct of fair criminal justice proceedings, and the transparency and accountability of security and justice providers, need considerable time and efforts before they become entrenched in traditional systems.

Community safety initiatives focusing on consultation and 'partnership' between formal and informal security and justice providers give crucial support in transitional periods. Practical steps for achieving this sort of integration can include the following.

- Providing a space for public dialogue and discussion about prevailing security policies, the roles and responsibilities of state security providers, and the vision for national security sponsored by the authorities.
- Providing a 'learning site' (Reich 2006: 24) in which key elements of the constitutional and legal framework for security and their implications for local constituencies are presented as valuable and viable tools.
- Offering the opportunity for local community representatives to discuss and review these laws and informing the overall security sector legal reform process.
- Providing space for law enforcement agents and traditional justice providers to exchange views on their roles within the society, and particularly as custodians of social peace, as a powerful device for promoting closer integration.
- Promoting the integration of local community safety structures into official by-laws, such as the organizational charts of governorate offices, ministries, and/or security agencies.

Fostering institutional changes to integrate community safety structures

Community safety initiatives need institutional anchorage in state structures (Homel 2009a; Idriss *et al.* 2010). While they can provide an inspiring model for the nationwide SSR process, local community safety successes might also antagonize central-level actors who were not involved in their development and implementation. Although the buy-in of local actors into top-down SSR processes is key, central-level decision-makers, politicians, and security leaders also need convincing about the relevance of community-based security initiatives.

Adopting a local community crime prevention approach to community safety initiatives can foster important institutional changes in the following ways.

- Promoting the adoption of organizational changes within state institutions responsible for decentralization or local security governance (ministries of the interior, local governance, and justice, the police, etc.).
- Creating community safety structures within a local administration (for instance the governors' offices).
- Developing capacity for collecting baseline data on crime and insecurity among the police and other law enforcement agencies.

- Developing the capacity to provide communities with a method and structure for developing formal citizens' complaints mechanisms by which abuses by formal and informal security providers can be documented and followed up. This is an important capacity that the crime prevention planning process can bring to bear in unstable environments with poor rule-of-law structure. In the case of Jenin, this proved to be a strong device for strengthening trust and promoting confidence that institutional change within the structures of the security environment was possible.

Local collaborative partnerships and the international context

The importance of the collaborative, partnership, and bottom-up methods illustrated by the strategic approach adopted in Jenin is reinforced by key sections of the Doha Declaration arising from the 13th UN Congress on Crime Prevention and Criminal Justice (UNODC 2015). The Doha Declaration (para. 10) asserts that all UN member states

> ... support the development and implementation of consultative and participatory processes in crime prevention and criminal justice in order to engage all members of society, including those at risk of crime and victimization, to make our prevention efforts more effective and to galvanize public trust and confidence in criminal justice systems.

In particular, the Doha Declaration (ibid.) urges all member states, among other measures:

> To plan and implement comprehensive policies and programmes that foster socioeconomic development, with a focus on the prevention of crime, including urban crime, and violence, and to support other Member States in such endeavours, in particular through the exchange of experience and relevant information on policies and programmes that have been successful in reducing crime and violence through social policies ... and to promote a culture of lawfulness based on the protection of human rights and the rule of law while respecting cultural identity, with particular emphasis on children and youth, seeking the support of civil society and intensifying our prevention efforts and measures targeting and using the full potential of families, schools, religious and cultural institutions, community organizations and the private sector in order to address the social and economic root causes of crime.
>
> (UNODC 2005: para. 10a and 10c)

While these observations and recommendations are of concern to the international community as a whole, there are key elements of the Doha Declaration that emphasize current challenges for crime prevention. These include the need to ensure that crime prevention policies and programmes respect and respond to the concerns of alienated or marginalized groups within the community, including 'vulnerable members of society, regardless of their status, who may be subject to multiple and aggravated forms of discrimination, and to prevent and counter crime motivated by intolerance or discrimination of any kind' (ibid.: para. 5).

A further highly pertinent challenge highlighted by the Doha Declaration is the need to 'integrate crime prevention and criminal justice strategies into all relevant social and economic policies and programmes, in particular those affecting youth, with a special emphasis on programmes focused on increasing educational and employment opportunities for youth and young adults' (ibid.: para. 7c). This point draws particular attention to the challenge of balancing 'top-down' and 'bottom-up' approaches to the prevention of crime (Homel and Masson 2016), a challenge that inevitably leads to considerations of how power and control are distributed in the task of creating safer communities (Homel and Homel 2012).

Conclusion

The conduct of effective SSR in fragile contexts requires the identification of key intersections with locally led community safety programmes. Similarly, bottom-up security initiatives are unlikely to be sustainable if they are not anchored in a nationwide security transformation process. Crucial intersection areas between bottom-up and top-down security initiatives can be identified in individual, legal, and institutional change processes.

The experience in Jenin demonstrated that when implemented well, a bottom-up community-based approach to crime prevention can provide significant advantages and strengths. The strategic crime prevention planning process can help to identify and map the security needs and concerns of citizens while at the same time providing a practical approach to involving local stakeholders in shaping decisions for their security.

What the use of modern local community-based crime prevention models shows is that local stakeholders are often in need of external support, technical expertise, and funding for developing, implementing, monitoring, and evaluating community safety plans (Homel 2009b). In unstable as well as in established countries, strategic partnerships between community-level security actors and state-level decision-makers are crucial for effectively bridging the gaps that define people's insecurity.

The Jenin experience demonstrated that the process of joining the top-down and bottom-up approaches to security through the adoption of key problem-solving concepts and techniques of local community crime prevention in combination with a modified SSR model may well provide an important alternative pathway to increased community safety, where the context is amenable.

More importantly, the collaborative approach and bottom-up model derived from the Jenin experience as outlined here resonates strongly with key elements of the international agenda for development of safe and sustainable communities. The 2030 Agenda for Sustainable Development (Goal 16) recognizes that reducing conflict, crime, violence, and discrimination and ensuring the rule of law, inclusion, and good governance are key elements of people's well-being and essential for securing sustainable development. Sustainable Development Goal 11 also explicitly highlights the promotion of safe, inclusive, and resilient cities, arguing that this must be achieved through equitable development safeguarded by fair, human, and effective crime prevention and criminal justice systems as a central component of the rule of law (UNGA 2015).

Underlying all this work is a need to recognize realistically and respond to the fact that in the absence of at least a minimal level of willingness to cooperate and work together to produce a safer community, no amount of good strategic planning and well-implemented programmes can hope to succeed. This is illustrated by the pressures faced by the OPT's local governorates in a continously tense political climate.

Acknowledgements

The views expressed in this chapter are those of the authors alone and do not necessarily reflect those of the Swiss Federal Department of Foreign Affairs or of DCAF.

References

AIC (Australian Institute of Criminology). 2003. 'Preventing Repeat Victimisation', AICrime Reduction Matters No. 12. Canberra, CT: AIC, available at http://aic.gov.au/publications/crm/crm012 (accessed 7 July 2018).

Baker, B. 2010a. 'The Future is Non-State', in M. Sedra (ed.) *The Future of Security Sector Reform*. Waterloo, ON: Centre for International Governance Innovation, 208–228.

Baker, B. 2010b. 'Linking State and Non-State Security and Justice', *Development Policy Review* 28(5), 597–616.

Ball, N. 2010. 'The Evolution of the Security Sector Reform Agenda', in M. Sedra (ed.) *The Future of Security Sector Reform*. Waterloo, ON: Centre for International Governance Innovation, 29–44.

Bronner, E. 2008. 'A West Bank Ruin, Reborn as a Peace Beacon', *New York Times*, 11 September, available at www.nytimes.com/2008/09/12/world/middleeast/12jenin.html (accessed 2 July 2018).

Bryden, A. and H. Hänggi (eds). 2005. *Security Governance in Post-Conflict Peacebuilding*. Geneva: DCAF.

Byrd, W. A. 2010. 'The Financial Dimension of Security Sector Reform', in M. Sedra (ed.) *The Future of Security Sector Reform*. Waterloo, ON: Centre for International Governance Innovation, 301–313.

Cherney, A. 2017. 'Designing and Implementing Programmes to Tackle Radicalization and Violent Extremism: Lessons from Criminology', *Dynamics of Asymmetric Conflict* 9(1–3), 82–94.

Cocayne, J. 2013. 'Strengthening Mediation to Deal with Criminal Agendas', Oslo Forum Papers 2. Geneva: Centre for Humanitarian Dialogue.

DCAF (Geneva Centre for the Democratic Control of Armed Forces). 2009. 'Palestinian Security Sector Governance: The View of Civil Society in Jenin', DCAF Spotlight 3. Geneva and Ramallah: DCAF, available at www.dcaf.ch/palestinian-security-sector-governance-view-civil-society-jenin (accessed 7 July 2018).

DCAF (Geneva Centre for the Democratic Control of Armed Forces). 2010a. 'Building the Rule of Law in Jenin Governorate: Interview with Dr. Qaddoura Mousa, Governor of Jenin', DCAF Spotlight 9. Geneva and Ramallah: DCAF, available at www.dcaf.ch/building-rule-law-jenin-governorate-interview-dr-qaddoura-mousa-governor-jenin (accessed 7 July 2018).

DCAF (Geneva Centre for the Democratic Control of Armed Forces). 2010b. 'Palestinian Women and Girls and the Provision of Security', DCAF Spotlight 8. Geneva and Ramallah: DCAF, available at www.dcaf.ch/palestinian-women-and-girls-and-provision-security (accessed 7 July 2018).

DCAF (Geneva Centre for the Democratic Control of Armed Forces). 2012. 'Developing a Community Safety Plan for Jenin.' Geneva and Ramallah: DCAF, available at www.dcaf.ch/sites/default/files/publications/documents/Developing-a-CS-Plan-for-Jenin_EN.pdf (accessed 7 July 2018).

Donais, T. 2009. 'Inclusion or Exclusion? Local Ownership and Security Sector Reform', *Studies in Social Justice* 3(1), 117–131.

ECOSOC (UN Council for Economic and Social Development). 2002. *Guidelines for the Prevention of Urban Crime (Resolution 1995/9, Annex) and the Guidelines for the Prevention of Crime (Resolution 2002/13, Annex).* New York: ECOSOC.

Edgar, L., C. Marshall, and M. Bassett. 2006. *Partnerships: Putting Good Governance Principles in Practice*. Ottawa, ON: Institute of Governance.

Friedrich, R. and A. Luethold. 2009. 'And They Came In and Took Possession of Reform. Ownership in Palestinian SSR', in T. Donais (ed.) *Local Ownership and Security Sector Reform*. Geneva: DCAF, 191–213, available at www.dcaf.ch/local-ownership-and-security-sector-reform (accessed 7 July 2018).

Friedrich, R., A. Luethold, and F. Milhem (eds). 2008. *The Security Sector Legislation of the Palestinian National Authority*. Geneva: DCAF, available at www.dcaf.ch/security-sector-legislation-palestinian-national-authority (accessed 7 July 2018).

Gordon, E. 2014. 'Security Sector Reform, Local Ownership and Community Engagement', *Stability: International Journal of Security and Development* 3(1), 1–18.

Hänggi, H. and V. Scherrer. 2008. 'Towards an Integrated Security Sector Reform Approach in UN Peace Operations', *International Peacekeeping* 15(4), 486–500.

Harzallah, N. and N. Masson. 2014. 'Relever les Défis de Sécurité Humaine dans le Nord-Ouest Tunisien', Regional Conference Report. Geneva and Tunis: DCAF, available at www.dcaf-tunisie.org//adminDcaf/upload/ejournal/documentfr_10169.pdf (accessed 2 July 2018).

Homel, P. 2005. 'A Short History of Crime Prevention in Australia', *Canadian Journal of Criminology and Criminal Justice* 47(2), 355–368.

Homel, P. 2009a. 'Lessons for Canadian Crime Prevention from Recent International Experience', *Institute for the Prevention of Crime Review* 3, 13–39.

Homel, P. 2009b. 'Improving Crime Prevention Knowledge and Practice', Trends and Issues in Crime and Criminal Justice 385. Canberra, CT: Australian Institute of Criminology, available at www.aic.gov.au/publications/current%20series/tandi/381-400/tandi385.html (accessed 2 July 2018).

Homel, P. and G. Fuller. 2015. 'Understanding the Local Government Role in Crime Prevention', Trends and Issues in Crime and Criminal Justice 505. Canberra, CT:

Australian Institute of Criminology, available at http://aic.gov.au/publications/tandi/tandi505 (accessed 7 July 2018).

Homel, P. and N. Masson. 2016. 'Partnerships for Human Security in Fragile Contexts: Where Community Safety and Security Sector Reform Intersect', *Australian Journal of International Affairs* 70(3), 311–327.

Homel, R. and P. Homel. 2012. 'Implementing Crime Prevention: Good Governance and a Science of Implementation', in B. Welsh and D. P. Farrington (eds) *The Oxford Handbook of Crime Prevention*. Oxford: Oxford University Press, 423–445.

Hulme, S. and P. Homel. 2015. 'Evaluation of the Victorian Community Crime Prevention Program: Final Report', special report. Canberra, CT: Australian Institute of Criminology, available at: http://aic.gov.au/publications/special/008 (accessed 7 July 2018).

ICG (International Crisis Group). 2010. 'Squaring the Circle: Palestinian Security Reform under Occupation', Middle East Report 98. Ramallah, Jerusalem, and Brussels: ICG.

Idriss, M., M. Jendly, J. Karn, and M. Mulone 2010. 'International Report on Crime Prevention and Community Safety: Trends and Perspectives, 2010.' Montreal, QC: International Centre for Prevention of Crime, available at www.crime-prevention-intl.org/fileadmin/user_upload/Publications/Crime_Prevention_and_Community_Safety_ANG.pdf (accessed 2 July 2018).

IPC (Institute for the Prevention of Crime). 2008. *What is Crime Prevention?* Ottawa, ON: University of Ottawa.

Jackson, P. 2011. 'Security Sector Reform and State Building', *Third World Quarterly* 32(10), 1803–1822.

Kritz, N. 2007. 'The Rule of Law in Conflict Management?', in C. Crocker, F. O. Hampson, and P. All (eds) *Leashing the Dogs of War: Conflict Management in a Divided World*. Washington, DC: US Institute for Peace, 401–424.

Larzillière, P. 2012. 'Production of Norms and Securitization in Development Policies: From Human Security to Security Sector Reform'. Beyrouth: American University of Beyrouth, available at www.aub.edu.lb/ifi/international_affairs/unaw/Documents/working_paper_series/20130110ifi_wps_unaw_Lazilliere.pdf (accessed 2 July 2018).

Mac Ginty, R. 2010. 'Hybrid Peace: The Interaction between Top-Down and Bottom-Up Peace', *Security Dialogue* 41(4), 391–412.

Mac Ginty, R. 2015. 'Where is the Local? Critical Localism and Peacebuilding', *Third World Quarterly* 36(5), 840–856.

Mannitz, S. 2014. 'From Paternalism to Facilitation: SSR Shortcomings and the Potential of Social Anthropological Perspectives', *International Peacekeeping* 21(2), 269–285.

Marten, M. 2014. 'Reformed or Deformed? Patronage Politics, International Influence, and the Palestinian Authority Security Forces', *International Peacekeeping* 21(2), 180–197.

Mobek, E. 2010. 'Security Sector Reform and the Challenges of Ownership', in M. Sedra (ed.) *The Future of Security Sector Reform*. Waterloo, ON: Centre for International Governance Innovation, 230–243.

Morgan, A. and P. Homel. 2011. 'A Model Performance Framework for Community-based Crime Prevention', Technical and Background Paper Series 40. Canberra, CT: Australia Institute of Criminology, available at http://aic.gov.au/publications/tbp/tbp040 (accessed 7 July 2018).

Morgan, A., H. Boxall, K. Lindeman, and J. Anderson. 2012. 'Effective Crime Prevention Interventions for Implementation by Local Government', Research and Public Policy Series 120. Canberra, CT: Australian Institute of Criminology.

Nathan, L. 2006. *Local Ownership in Security Sector Reform. A Guide for Donors*. London: Department for International Development.

OECD (Organisation for Economic Co-operation and Development). 2007. *OECD-DAC Handbook on Security System Reform: Supporting Security and Justice*. Paris: OECD Publishing.

OECD (Organisation for Economic Co-operation and Development). 2008. 'The Paris Declaration on Aid Effectiveness and the Accra Agenda for Action'. Paris: OECD, available at www.oecd.org/dac/effectiveness/34428351.pdf (accessed 7 July 2018).

PCI (Peaceful Change Initiative). 2013. 'A Peace Building Agenda for Libya'. London: PCI, available at www.peacefulchange.org/uploads/1/2/2/7/12276601/2012_10_a_peacebuilding_agenda_for_libya_eng.pdf (accessed 7 July 2018).

Redo, S. 2012. *Blue Criminology. The Power of the United Nations Ideas to Counter Crime Globally*. Helsinki: European Institute for Crime Prevention and Control.

Reich, H. 2006. '"Local Ownership" in Conflict Transformation Projects: Partnership, Participation or Patronage?', Occasional Paper 7. Berlin: Berghof Foundation.

Richmond, O. P. 2012. 'Beyond Local Ownership in the Architecture of International Peacebuilding', *Ethnopolitics* 11(4), 354–375.

Saferworld. 2013. *Community-Based Approaches to Safety and Security. Lessons from Kosovo, Nepal and Bangladesh*. London: Saferworld.

Sampson, R. J. 2013. 'The Place of Context: A Theory and Strategy for Criminology's Hard Problems', *Criminology* 51(1), 1–31.

Sedra, M. 2006. 'European Approaches to Security Sector Reform: Examining Trends Through the Lens of Afghanistan', *European Security* 15(3), 323–338.

Sedra, M. (ed.) 2010. *The Future of Security Sector Reform*. Waterloo, ON: Centre for International Governance Innovation.

Seyigh, Y. 2011. 'Policing the People, Building the State: Authoritarian Transformation in the West Bank and Gaza', Carnegie Papers. Washington, DC: Carnegie Endowment for International Peace, available at http://carnegieendowment.org/files/gaza_west_bank_security.pdf (accessed 2 July 2018).

UNDP (UN Development Programme). 1994. *Human Development Report*. New York: UNDP.

UNDPKO (UN Department of Peacekeeping). 2012. *The United Nations SSR Perspectives*. New York: UNDPKO, available at www.un.org/en/peacekeeping/publications/ssr/ssr_perspective_2012.pdf (accessed 2 July 2018).

UNGA (UN General Assembly). 2014. 'The Rule of Law, Crime Prevention and Criminal Justice in the United Nations Development Agenda Beyond 2015', UN Doc. A/RES/68/188, 11 February. New York: United Nations, available at www.un.org/en/ga/search/view_doc.asp?symbol=A/RES/68/188&referer=www.un.org/en/ga/68/resolutions.shtml&Lang=E (accessed 2 July 2018).

UNGA (UN General Assembly). 2015. 'Transforming Our World: The 2030 Agenda for Sustainable Development', UN Doc. A/RES/70/1, 21 October. New York: United Nations, available at www.un.org/ga/search/view_doc.asp?symbol=A/RES/70/1&Lang=E (accessed 2 July 2018).

United Nations. 2014. *Global Governance and Global Rules for Development in the Post-2015 Era*. New York: United Nations.

UNODC (UN Office on Drugs and Crime). 2010. *Handbook on the Crime Prevention Guidelines: Making Them Work*. Vienna: UNODC, available at www.unodc.org/documents/justice-and-prison-reform/crimeprevention/10-52410_Guidelines_eBook.pdf (accessed 2 July 2018).

UNODC (UN Office on Drugs and Crime). 2015. 'Doha Declaration on Integrating Crime Prevention and Criminal Justice into the Wider United Nations Agenda to Address Social and Economic Challenges and to Promote the Rule of Law at the National and International Levels, and Public Participation', UN Doc. A/CONF.222/L.6, 31 March. Vienna: UNODC, available at www.unodc.org/documents/congress//Documentation/IN_SESSION/ACONF222_L6_e_V1502120.pdf (accessed 2 July 2018).

UNSSR Inter-Agency Task Force. 2012. *Security Sector Reform Integrated Technical Guidance Notes.* New York: Office of the UN Secretary-General.

van Dijk, J. and J. de Waard. 1991. 'A Two Dimensional Typology of Crime Prevention Projects: With a Bibliography', *Criminal Justice Abstracts* 23, 483–503.

Welsh, B. and D. P. Farrington 2012. 'Crime Prevention and Public Policy', in B. Welsh and D. P. Farrington (eds) *The Oxford Handbook of Crime Prevention.* Oxford: Oxford University Press, 3–22.

8 Safe services in unsafe cities

The ICRC and urban services in situations of armed conflict and urban violence

Hugo Slim, Caroline Putman-Cramer, Evaristo De Pinho Oliveira, and Rachel Banfield

Introduction

War between cities and violence in cities are not just modern phenomena. The first significant wave of urbanization in human history took place in the fifth millennium BCE in the fertile plains of Mesopotamia, or what is now called the Middle East (Mark 2014). As people turned from pastoralism to agriculture and developed irrigation systems, they settled in cities and began to develop urban life. The great cities of the ancient Middle East were the first flourishing of sophisticated human settlement where grand ambitions for architecture, planning, services, and urban governance were put into practice. Defended walled cities were common by 2900 BCE, and trade between these ancient cities was accompanied by war, conquest, the imposition of empire, and the issue of government.

It was in these Mesopotamian cities that people developed writing for the first time, and literature began in cities too. Interestingly, and perhaps not surprisingly, the earliest human story available in written form comes from these cities and begins with the problem of urban violence. The *Epic of Gilgamesh* tells the story of Gilgamesh, the great king of Uruk who is two-thirds god and one-third man (George 1999). Gilgamesh is out of control. He rules his city tyrannically by fighting and oppressing the city's young men and insisting on his sexual right to the city's young women on the night of their marriage. The people of Uruk cry out to their gods to save them from the violence of Gilgamesh's rule. In response the gods make Enkidu, an equal to Gilgamesh who challenges, befriends, and diverts him in a quest for enlightenment. Eventually, having seen his friend die and searched in vain for immortality, Gilgamesh returns to Uruk a wiser man who realizes that the best form of eternity open to humankind is the just and sustainable rule of a city from one generation to the next. The good city is as good as we get as human beings.

But it remains as hard as ever to make good cities, and the challenges of conflict, violence, and urban government have continued throughout history. After Mesopotamian civilization these problems were taken up anew in Greek philosophy: the word politics is taken directly from the Greek *polis*, meaning city. Politics developed as a domestic challenge of right relationships *within* the city and as an international challenge of right relationships *between* city states and competing empires.

Today, urban warfare and violence are again at the top of political and humanitarian agendas – and ancient and modern cities in the Middle East are once again at the centre of armed conflicts. In many of the world's small, medium-sized, and mega-sized cities interpersonal and collective violence in various forms is ruining the lives and life chances of many millions of people, just as it did in Uruk thousands of years ago until Gilgamesh became enlightened. Most people still settle in cities in the hope of making a better life for themselves and their families. They hope that urban life will bring them closer to a decent livelihood, better services, and a thriving communal life. Cities deliver on these hopes for many millions of people, but not in conditions of armed conflict and chronic urban violence.

The urban operations of the International Committee of the Red Cross (ICRC) have started to rise sharply in the last 10 years, and humanitarian action more generally is in the midst of an urban revolution as its practices change dramatically to cope with urban crisis (Campbell 2016). This chapter looks at the expanding urban work of the ICRC and sets out a practical humanitarian agenda for improving people's safety in cities affected by armed conflict and chronic urban violence. The chapter draws on the ICRC's operational experience in armed conflict in the Middle East, Africa, and Asia and its work in settings of urban violence in Latin America. It examines some of the dynamics of danger and risk for vulnerable populations in different urban areas, and illustrates how the ICRC and its partners in local authorities and national Red Cross and Red Crescent Societies work to limit these risks and maintain essential services. The chapter describes the general approach of the ICRC, and then looks in detail at the specific challenges of supporting safe services in armed conflict and urban violence, such as safe water, safe health facilities, and safe schools.

Safe access to urban services

The population density in cities, towns, and informal urban settlements means that people's needs must be met at a scale that is different from rural areas. This makes sophisticated and interconnected urban services for water, electricity, sewerage, healthcare, and education the main focus for humanitarian response. Faced with armed conflict and violence, service continuity and people's access to these services become essential to survival and a life with dignity.

The principle of safe access to essential services is a constant theme in the United Nations (UN) New Urban Agenda (NUA) and is rightly prioritized in settings of armed conflict and urban violence, just as it is in less conflictual urban settings (UN-Habitat 2016: paras 40, 50, and 99). Safety is a necessarily broad concept in urban policy and practice, and one which the ICRC usually addresses as 'protection', linking it directly to legal protections and the obligations of authorities and rights of individuals in international and domestic law. Safety and protection have many dimensions. For example, safe access to water requires people's safe and unrestricted movement to and from water sources, as well as proper quality levels to ensure the water is safe to drink. In the same way, people should be able to reach health centres and schools safely, and then also be safe inside these vital facilities.

The need for safer cities is rightly emphasized today as a global policy priority. Reducing violence is an important part of making cities safer, but so too is expanding people's inclusion in urban life by improving their safe access to essential services and minimizing the disruption of these services. Nowhere is this challenge greater than in armed conflicts and in the many urban areas of the world affected by chronic violence.

Cities, violence, and armed conflict today

In an urbanizing world, armed conflict and violence are urbanizing too. War and violence are entrenched features across today's urban landscape. This fact is well recognized in the NUA, and its call for action insists that 'special attention should also be given to countries in situations of conflicts, as well as countries and territories under foreign occupation, and post conflict countries' (ibid.: para. 19). The NUA also makes special note of the need to address the multiple forms of discrimination faced by many different groups, including internally displaced people (IDPs), refugees, and migrants (ibid.: paras 20 and 42). This is often especially important in their access to land, services, and employment – a policy point consistently emphasized by the Norwegian Refugee Council and the International Federation of Red Cross and Red Crescent Societies (NRC and IFRC 2016).

Armed conflicts over the last decade have destroyed many towns and cities in the Middle East, as well as in Ukraine, Somalia, South Sudan, the Central African Republic, Nigeria, and elsewhere. War has also created new towns and cities by forced displacement and its brutal changes to demographic patterns. Numbers of IDPs, migrants, and refugees are expanding and transforming the urban space of small towns and large cities. Armed conflict is driving urbanization in cities like Maidiguri in Nigeria and Mogadishu in Somalia. Large humanitarian 'camps' of people who have been forcibly displaced for decades in places like Dadaab in Kenya and Abu Shoak in Darfur are now de facto towns (Osman Ali and Mahmoud 2016). Permanent urbanization often follows apparently temporary flight and encampment. In the fifth century, Venice was founded by displaced people fleeing for their lives from successive waves of Germanic and Hun invasions into today's northern Italy. Thousands of displaced people hid in the marshes bordering the Adriatic and gradually settled on the various islands of its lagoon to form what was to become one of the most beautiful and influential cities in the world.

Not all violence in cities is armed conflict. Much chronic urban violence involves the activities of gangs, drug traders, and other criminal armed groups (see also Chapter 4 in this volume). The Small Arms Survey estimated that on average 377,000 people are killed each year by intentional homicide (Geneva Declaration Secretariat 2015), and a large proportion of these deaths occur in urban areas, with rates being of 'epidemic proportions' in some Latin American and Caribbean countries (World Bank 2016). The particular violence of 'femicide' – the deliberate killing of women – is also a growing phenomenon in urban areas across parts of Latin America, and is often accompanied by sexual violence and rape (WHO 2012).

The extent of urban violence and the humanitarian consequences of this violence on people's lives, dignity, and safe access to urban services are becoming a major concern for the ICRC. In the last few years the ICRC has been working with municipal authorities, weapons bearers, and affected communities in several cities in Latin America, Asia, and Africa to try to reduce the suffering caused by urban violence and increase people's protection.

The ICRC's operational approach

The ICRC is an impartial, neutral, and independent organization whose exclusively humanitarian mission is to protect the lives and dignity of victims of armed conflict and other situations of violence and provide them with assistance. The ICRC also endeavours to prevent suffering by promoting and strengthening international humanitarian law and universal humanitarian principles. Established in Geneva in 1863, the ICRC is at the origin of the Geneva Conventions and the International Red Cross and Red Crescent Movement. It directs and coordinates the international activities conducted by the overall movement in armed conflicts and other situations of violence.

The ICRC works to protect people from armed violence in some of the most dangerous cities in the world. Its strict adherence to humanitarian principles means it works across the lines of conflict and with all parties to conflict. Its focus on preventing human suffering and protecting and assisting people means the ICRC is involved in a wide range of activities in urban areas today. It does this in close operational partnerships with national Red Cross and Red Crescent Societies and their exceptional networks of volunteers at the community level. Together, they work with local authorities of all kinds to maintain basic services including water, sanitation, energy, health, and education. They also work with vulnerable groups to improve their shelter, food security, and household income.

This humanitarian assistance is constantly accompanied by the ICRC's efforts to protect people from the risk of violence and other abuses. Its law and protection work involves engaging with all weapons bearers to ensure greater respect for international humanitarian law (IHL) in armed conflicts, and for the relevant standards of international human rights law (IHRL) in other situations of violence. The ICRC's protection activities often see it working directly with affected communities and paying particular attention to the needs of the most vulnerable, including people in detention, victims of sexual violence, urban IDPs, vulnerable migrants, and their host communities. The ICRC and the wider Red Cross and Red Crescent Movement also operate an extensive global network that searches for missing people and reconnects family members who have become separated as a result of conflict, armed violence, or migration.

Urban services and armed conflict

There are important differences and some similarities between urban armed conflict – often referred to as urban warfare – and chronic urban violence. Both types of violence happen in towns and cities. Both may be protracted and last for many years, and each form of violence may have widespread and devastating humanitarian consequences for the people and places affected. For example, urban warfare and chronic urban violence may force people to flee their homes. But the scale and types of weaponry in urban warfare can be dramatically different to those used in urban violence, as illustrated below. One very significant difference is the applicable legal framework for each context: while IHRL and national legislation in principle apply in both urban armed conflicts and situations of armed violence, IHL is only applicable in – and is the relevant legal framework for – situations of armed conflict.

People's safe access to urban services is recognized in the various rules of IHL and has become a major priority for the ICRC, which has scaled up in the last 10 years to meet the urban challenge. ICRC capacity has been transformed from traditional rural practice focused on water, sanitation, electricity, and health services for villages and IDP camps to strategic supply and service continuity for large urban populations of millions of people. Urban warfare in today's towns and cities frequently destroys or degrades infrastructure and services that are essential to the safety and survival of the civilian population (ICRC 2015). Two particular challenges are a constant feature of the humanitarian struggle to protect and assist civilian populations in urban warfare: the cumulative impact of protracted conflict, and the need for greater respect for IHL in the way urban wars are fought.

The cumulative impact of protracted conflict

Most armed conflicts today are protracted conflicts lasting for decades – like in Iraq, Afghanistan, South Sudan, the Democratic Republic of Congo, and Israel and the Occupied Palestinian Territories (ICRC 2016a). Even in conflicts that have lasted for years rather than decades, as in Syria, Yemen, the Central African Republic, and northern Nigeria, the levels of destruction and degradation can run deep in infrastructure and basic services, some of which may not have been very good even before the onset of hostilities. The ICRC's 2015 study, *Urban Services During Protracted Armed Conflict*, analyses the impact of armed conflict on essential urban services (ICRC 2015). Its findings make a strong legal and humanitarian case for prioritizing support to the continuity of essential services to protect and assist vulnerable civilian populations. The Geneva Conventions make clear that civilian objects 'indispensable to the survival of the civilian population' must be protected, and that water, sanitation, healthcare, and electricity are essential to people's survival alongside food and shelter (ibid.).

People in urban areas rely on an intricate network of essential services to survive and thrive. Power stations supply electricity through long networks of overhead cables and substations. This electricity powers pumping stations that distribute water along underground pipes and store it in reservoirs and tanks of various kinds. Electricity and water are both vital to maintain healthcare facilities, schools, and homes. The growing sophistication of urban infrastructure means that services like electricity, water, sewerage, waste disposal, and healthcare rely more than ever on complex logistics and specialist personnel. Urban residents are highly dependent on these complex essential services and usually do not have the means to restore the services themselves once they are damaged or run down. The scale and interconnectedness of services and infrastructure are such that the disruption of one essential service during conflict can have major reverberating effects. One electricity substation out of action can mean power and water failures for hospitals, schools, businesses, and homes for several weeks. This kind of break in the chain of interconnected services can affect many thousands of people in different ways, including having no drinking water, no power for cooking, lighting, or business, and interrupting health and education provision. In armed conflicts these breaks in the chain happen more often and tend to last longer because of a decline in resources to fix them.

Over time armed conflicts can have a seriously damaging cumulative impact on urban services, leading to long-term deterioration in their capacity, performance, and effectiveness, with potentially severe consequences for people's dignity, health, and long-term life chances. The ICRC's 2015 report highlights a conflict's devastating impact on the 'critical people, critical hardware and critical consumables' that are essential to the efficient running of complex urban services. It notes how various factors combine to cause this inevitable deterioration of urban services in protracted conflicts: repeated damage to essential civilian infrastructure; difficulty in obtaining the necessary spare parts and consumables; the 'brain drain' of skilled staff who are displaced or killed; and the obstacles to maintaining infrastructure over contested front lines (ibid.: 21–31). This impact on urban services is not limited to places where the fighting is taking place. Urban systems – particularly those that are already fragile – are also strained to breaking point in cities that have to host large numbers of displaced people with dwindling municipal budgets and rising costs.

The ICRC has been in Iraq since 1980, responding to the humanitarian needs that have arisen as a result of decades of changing conflict. It has become increasingly creative in adapting its work to meet short-term needs while also supporting the services that are critical to people's lives over the medium and long terms. For 25 years the ICRC has directly supported local service providers in repairing and rehabilitating water supply infrastructure. One example of this short–long combination was in 2015 in the Al Ramadi camp, Al Amal camp, and the informal settlements around the Caravan Camp in Anbar. There, the ICRC ensured immediate access to water for IDPs by installing water tanks and tap stands, and trucking water from nearby towns. It simultaneously rehabilitated the water supply infrastructure in nearby towns in coordination with the Khalidiyah Directorate of Water to ensure the expanded supply required by the needs of the IDPs. This provided a more reliable supply of water for approximately 21,900 local residents in these areas, as well as securing the water supply for approximately 37,000 IDPs in the camps (ICRC 2016a: 29).

One particular objective of the ICRC's urban work is to minimize the cumulative impact of conflict on civilians and develop significant expertise and capacity in support for strengthening the resilience of municipal service delivered during armed conflict and situations of extreme fragility. About 60 per cent of the 2016 budget of the ICRC's Water and Habitat Unit was invested in urban areas, most of it to maintain water supply networks (ICRC 2016c: 4). New expertise has enabled the ICRC to maintain large-scale electricity, water, sewerage, and waste disposal systems in many cities, towns, and informal settlements affected by armed conflict. This 'shift to urban' in the ICRC's water and sanitation work has been a deliberate move in the last ten years and involved a veritable revolution in skills, scale, and expertise. In the largely rural armed conflicts of previous decades, major urban interventions were the exception. Today they are the norm. The ICRC employs electrical engineers specialized in high voltage electricity supply and urban water and sewerage engineers, and partners with municipal authorities on an unprecedented scale.

ICRC teams bolster the resilience of complex interconnected systems in the face of shocks and supply shortages that make these systems so vulnerable during armed conflicts. This means working in close operational partnerships with water authorities and local commercial contractors who know the systems and infrastructure well. The ICRC helps to ensure service continuity by planning ahead, bringing in spare parts and skilled staff to plug gaps, and using the ICRC's principled and neutral approach to ensure that services continue on all sides of a conflict. This typically involves working across the lines of conflict, as explained by Hamed, a water technician from a water plant on the southeastern outskirts of Aleppo:

> During the crisis, we had to work in two water stations and used to cross the front lines on a weekly basis, in coordination with humanitarian partners and conflict parties on the ground. Our target was to keep the water facility working, and to ensure that water remained neutral.
>
> (ICRC 2017: 50)

During one high-risk mission to an area controlled by three different parties to the conflict, clashes broke out and Hamed was shot and had to be taken to hospital.

Since the outbreak of conflict in Syria, infrastructure that provides access to clean water has been subject to repeated damage and gradual deterioration because of inadequate operation and maintenance. Access to clean water in a country where some seven million people have been internally displaced – in large part to urban areas – has declined and in some areas is worryingly low. The ICRC began working early to respond to this critical mix of immediate and medium-term needs. Syria has always had a highly skilled workforce of engineers, and the ICRC drew on this local capacity to help support water systems in the most sustainable way. It has also given training on water, sanitation, and hygiene in emergencies to the volunteers of the Syrian Arab Red Crescent, and supported the hiring of Syrian engineers who work as front-line responders on immediate needs to complement the ICRC's emphasis on longer-term system resilience.

By investing in service continuity and working with local authorities across the lines of conflict, the ICRC and its partners in municipalities and public utilities have frequently been able to hold services at an effective level for large urban populations and prevent even greater suffering and development reversals. Maintenance of resilient infrastructure in the midst of conflict ensures the means of survival for many millions of people and preserves their dignity. It also makes a very basic contribution to the UN Sustainable Development Goals (SDGs) by preventing even greater impoverishment, and reduces the costs of post-conflict development. Humanitarian action of this kind secures 'development holds' by preventing the total collapse of urban infrastructure and services, and plays some part in the intermediate objectives of SDG 9 and SDG 11.

In Gaza the ICRC has been providing technical operational and financial support to the Coastal Municipalities Water Utility since 2006. To maintain service provision, the ICRC supports the utility in the rehabilitation of infrastructure and the supply of spare parts and consumables to enable more reliable operation and maintenance. During the 2014 war the ICRC acted as the sole point for coordination between the parties to the conflict, as part of a humanitarian effort to support the water, wastewater, and electrical service providers who were in need of safe access to perform repairs, basic operations, and maintenance on critical infrastructure. Again, the ICRC delivered a combined approach of short- and long-term work. It supported a short-term emergency response of water trucking and generators for a back-up power supply and, in parallel, rehabilitated damaged infrastructure.

This approach is not cheap, and financial support for this kind of sophisticated, large-scale infrastructure work does not fit easily into conventional categories of humanitarian and development funding. It can seem too structural and long term for some humanitarian budgets and too high risk and agile for development investors. Yet the maintenance of such large and complex systems over the duration of a protracted conflict requires something closer to extensive, well-planned, multiyear development financing than to the short-term funding typically required for humanitarian activities. International financial institutions and development agencies must lean into situations of conflict more pragmatically and finance urban services during protracted conflicts and situations of extreme fragility. This means development actors being willing to tolerate risk, and shaping funding mechanisms that are both agile and long term. The recent partnership between the World Bank and the ICRC in Somalia on water and health services is a step in the right direction and could signal a way for others to follow (World Bank 2017).

Greater respect for IHL in urban warfare

Supporting resilient urban services during armed conflict is an essential contribution to the protection and assistance of civilian populations in cities and towns, and has now been clearly recognized as sound humanitarian policy in the NUA. Paragraph 30 states that:

> We acknowledge the need for governments and civil society to support resilient urban services during armed conflicts. We also acknowledge the need to reaffirm full respect for international humanitarian law.
>
> (UN-Habitat 2016: para. 30)

The last sentence leads into the second major challenge in ensuring safe urban services for the civilian population in armed conflicts: the need for much greater respect of IHL, or the laws of war, in urban warfare. Some 50 million people are affected by armed conflict in urban areas today. Many of them have become urban IDPs. Ensuring the protection of these people, their host communities, and the urban services on which they all depend starts first and foremost with better respect for existing rules of IHL. The NUA reaffirms these rules.

IHL requires parties to armed conflicts to avoid – as much as possible – situating military objects within or near densely populated areas. This requirement is also relevant to state planning during peacetime. Urban planners should ensure that military objects are, whenever feasible, sited at a distance from populated areas to reduce the risk to civilians during any future armed conflict. Urban warfare also imposes major challenges on military commanders operating in accordance with IHL, as they have to take into account the vulnerability of large numbers of people living in densely populated urban areas and the intricacy and interconnectedness of essential services. The following principles and rules are central to urban warfare, as they are to all armed conflict. Respecting these rules helps to ensure better protection for civilians living in urban areas and reduces the incidence of urban displacement.

The principle of *distinction* requires parties to a conflict to distinguish between civilians and civilian objects on one hand and combatants and military objects on the other. In urban areas, military objects are often situated among persons and objects protected under IHL. This means that targeting often requires greater consideration in urban warfare than in open battlefields. Indiscriminate attacks – such as area bombardments – which make no distinction between military objects and civilians or civilian objects are prohibited.

The principle of *proportionality* is similarly important. Disproportionate attacks are prohibited – that is attacks which may be expected to cause excessive incidental harm to civilians in relation to the anticipated military advantage. The incidental harm that requires consideration is not limited to civilian death and injuries or damage to civilian objects caused directly by the attack; it includes indirect death, injury, or damage caused by the short- and long-term impact on infrastructure essential to the survival of the civilian population. The interdependence of urban systems means that the consequences of an attack are likely to spread far beyond the immediate vicinity of the impact zone, and affect large numbers of people in many different ways. For example, damage to one mains water pipe can dramatically reduce tens of thousands of people's access to water. All the reasonably forseeable effects of an attack – direct and indirect – must be taken into account by military commanders in their proportionality assessments.

The principle of *precaution* in attack requires that belligerents must take all feasible precautions to avoid – or at least minimize – incidental harm to civilians.

They must choose their means and methods of attack with this principle in mind, and must adapt them to the urban environment in which they are fighting.

Urban warfare in the Middle East has once again involved the widespread use of heavy artillery and aerial bombardment. These weapons systems raise significant issues of distinction, proportionality, and precaution, and the ICRC is particularly concerned about the use of explosive weapons with a wide impact area in densely populated urban areas. These kinds of weapons include large bombs and missiles; indirect-fire weapon systems like mortars, rockets, and artillery; multibarrel rocket launchers; and certain kinds of improvised explosive devices like so-called barrel bombs. In urban warfare combat takes place in neighbourhoods where fighters are often mixed in among civilians. Even when aimed at an awful target (i.e. military objective), explosive weapons with a wide impact area are likely to have indiscriminate effects, as blast and fragments fly through the air in all directions (ICRC 2017: 54).

In Syria, Iraq, Yemen, Ukraine, Afghanistan, Gaza, and elsewhere the ICRC has seen how the use of explosive weapons with a wide impact area in urban areas exposes civilians to great risk of incidental or indiscriminate death, injury, and lifelong disability. These attacks cause massive population displacement and immeasurable psychological trauma. Explosive weapons also have a significantly detrimental impact on infrastructure and essential services. Again, the interconnectedness of cities renders urban services vulnerable to a 'domino effect' of severe humanitarian consequences. This disruption often comes at a time when the services are needed most. The same blast that leads to a hospital being unable to provide effective services is also likely to cause large numbers of casualties who require hospital care. Even after armed conflict ends, long-lasting damage to infrastructure that provides essential services, or the presence of explosive remnants of war, can seriously hinder sustainable urban development. Since 2011, the ICRC is calling for all parties to conflict to avoid using explosive weapons with a wide impact area in densely populated areas.

Two other pressing issues of respect for IHL are critical in contemporary urban warfare: siege warfare, and the protection of the medical mission. Some of the most extreme suffering in armed conflicts in today's towns and cities is experienced by people living under siege. IHL sets out clear rules protecting civilians that are very relevant to siege. Many of these rules have not been respected in recent urban warfare, so starvation, family separation, and the collapse of health and water services have had severe humanitarian consequences for besieged populations.

The protection of health facilities, transport, personnel, and patients is also a major priority for people's safe access to urban services. Health services are often disrupted by violence during armed conflict or other emergencies: health-care personnel are attacked or threatened, patients are discriminated against and even deliberately killed, healthcare facilities are targeted, and medical supplies looted. These disruptions occur when healthcare services are needed most. The short-term impact of such violence is obvious: medical staff are prevented from providing critical assistance and patients are deprived of essential care. The long-term impact is equally serious: hard-won gains, such as improvements in maternal health or reductions of child mortality, can be erased in minutes, causing significant setbacks for sustainable development.

IHL and other applicable international legal norms aim to safeguard access to impartial healthcare. IHL, where applicable, protects medical personnel and their equipment and means of transport, and hospitals and other medical facilities, from violence. It also provides protection for the performance of medical activities in line with medical ethics. The global problem of attacks against medical missions was recognized by the UN Security Council in 2016 in its Resolution 2286. The resolution strongly condemns such attacks and demands that all parties to armed conflict fully comply with their obligations under international law and develop effective measures to protect medical facilities, transport, personnel, and patients (UNSC 2016).

Urban services and chronic urban violence

Beyond armed conflicts, many of the world's cities have areas experiencing chronic and serious armed violence. In many such cities, social and economic exclusion often combine with a retreat of state services to entrench poverty and vulnerability. Rising levels of crime and violence are frequently linked to the drugs trade, gangs, and the pursuit of particularly violent forms of masculine identity and status. Governmental response is typically framed in militarized models of law enforcement which engage the forces of violence with similarly forceful *mano dura* strategies of zero tolerance to create a highly confrontational environment (Koonings and Krujit 2015: 1–29).

The great majority of people who suffer the direct and indirect consequences of this violence are not the fast-growing middle classes of Latin America, Africa, and Asia but the urban poor who remain trapped in the most marginalized and high-risk areas (Auyero *et al.* 2015). These urban areas can have such high levels of chronic violence that it can be too risky just to go to school or attend a clinic. Even if people do reach these services, they may find that the staff are not there because of fears for their own safety or the services are too run down to function. Urban violence can have a deeply adverse impact on people's daily life – especially on their movements, their safety, and their well-being. In violent contexts, local authorities often find it difficult to provide essential public services without interruption. Investment in these areas is not always a priority, and the inclusion of people living in violent urban areas is a major challenge to national and local government authorities.

In its work in many urban areas across Latin America, the ICRC has seen the widespread and direct consequences of chronic urban violence: large numbers of people wounded and dead. The ICRC is also keenly aware of the less visible effects of urban violence, including forced displacement, sexual violence, disappearances, the recruitment of minors into armed groups, long-term psychosocial impact on people of all ages, and a lack of safe access to basic services. The World Bank noted the clear link between violence and poverty in its major study of violence and development:

> While much of the world has made rapid progress in reducing poverty in the last 60 years, areas characterized by repeated cycles of political and criminal violence are being left far behind, their economic growth compromised and their human indicators stagnant.
>
> (World Bank 2011: 1)

The ICRC sees the impact of chronic cycles of violence every day. The impact of violence on the provision of healthcare and education can be severe. People lose access to healthcare because of a fear of violence around the nearest health centre, the closure of health centres, the reluctance of health professionals to work in the areas affected, and the obstacles to outreach activities. All these affect the delivery of healthcare and reduce health indicators. Educational services are also affected by urban violence. In many areas where the ICRC works, incidents of armed violence put students and teachers at risk. Teachers can be threatened and students may be subject to recruitment into armed groups. Schools can be caught in the cross-fire, and school closures occur regularly. The psychological distress caused by close proximity to armed groups, and the behaviour of such groups towards a school community, can have broad-ranging consequences, including high rates of absenteeism, high staff turnover, and violent tensions within the school itself. In El Salvador up to 39,000 students dropped out of public schools due to violence in 2015, and in 2016 nine teachers and 71 students were violently killed (Cue and Nunez-Flores 2017).

A first step in all the ICRC's work is to alert the authorities to the need to measure the humanitarian impact of urban armed violence, and then support them in doing so – and this means understanding and quantifying the extent of the problem. To respond effectively to the consequences of armed violence in urban areas, a better understanding of its impact on people, public services, and development is needed as an evidence base. This requires all relevant government institutions to strengthen their capacity to collect and analyse reliable, representative, and disaggregated data. This includes the number of days that schools or health centres were closed because of violence, the number of cancelled family home visits by health workers, and the number of times emergency services were obstructed. These and many other indicators of the impact of urban violence need to be collected and used strategically in the design, planning, and monitoring of programmes to mitigate the effect of violence and improve people's safe access to services by addressing the causes as well as the symptoms of the problem.

The next step is to take a holistic approach to programming in violent areas. The ICRC's experience reveals a broad range of problems needing a wide range of creative interventions. The difficulty armed violence creates for service delivery often leads local authorities to take a mainly law enforcement and security approach. This is not enough. The ICRC recognizes the vital role law enforcement plays in these contexts, but it also believes that law enforcement can be effective only when it is part of a broader multidisciplinary response. A comprehensive approach to violence aims to empower both service professionals and affected communities. This means mobilizing and supporting municipal and national authorities across a broad front. The ICRC is able to work closely with national Red Cross and Red Crescent Societies and their existing grassroots networks as a major complement to government and community efforts.

One important objective in these partnerships is to build the confidence and skills of communities and service providers so they are better able to work safely and effectively in a violent context. This requires staff and communities to be adequately trained in particular ways of working to identify risks, minimize exposure, and develop security management protocols that can help to ensure the safety of service delivery agencies, their staff, and their clients. This is now a priority in the NUA, in which states affirm that 'we will take steps to include appropriate training and support for service delivery professionals and communities living in areas affected by urban violence' (UN-Habitat 2016: para. 99).

The ICRC's long experience of working with at-risk communities and authorities on protection strategies was brought together in its *Professional Standards for Protection Work* (ICRC 2013). This guidance document was created in consultation with numerous humanitarian and human rights actors, and can make a very useful contribution to increasing safety in cities. These standards highlight the importance of a meaningful dialogue with affected people about the threats they face, and how they can cope by identifying and developing self-protection strategies.

This community-centred approach is at the heart of all the ICRC's work in settings of urban violence, and is close to the new approach to urban resilience being developed *inter alia* by Koonings and Krujit (2015), with its focus on different 'modes of resilience' such as coping, mitigation, recovery, prevention, and transformation, working with a range of different actors. The ICRC's efforts in Rio de Janeiro are a case in point. In Rio, the ICRC realized that violence in certain *favelas* had a significant impact on the provision of public healthcare. Between November 2012 and October 2013 six health centres reported that over 25,000 home visits by community health workers and almost 3,000 clinical consultations had to be cancelled owing to armed violence. To lessen the risk of such violence to health professionals and improve their access to these areas of the city, the ICRC and the Municipal Secretariat for Health ran a training course as part of the ICRC's Safer Access programme (ICRC 2016b). The programme helps health professionals identify and classify the signs of imminent danger and determine what course of action would minimize the risk to their person. The programme was extended to all at-risk areas of the city and more than 2,500 health professionals had been trained by 2016. The results have been dramatic: primary healthcare services have since been provided to roughly 86 per cent of the population in violence-affected areas (ICRC 2016c: 10).

Community members must also be given support in self-protection as they seek to access services. This training and awareness raising mitigates the risk to their safety and prevents harmful coping strategies. The ICRC aims to ensure that individual rights are respected, and that the people affected are agents of their own protection when they seek to obtain basic services safely.

Young people are often a particular focus of safer behaviour programmes. In Ciudad Juárez, Mexico, the Ministry of Education, the Mexican Red Cross, and the ICRC are empowering young people to develop safer behaviour in their daily lives. The programme encompasses psychosocial support and the acquisition of life skills such as critical thinking, active listening, tolerance and empathy, conflict management, teamwork, and planning for one's future. This is supplemented by encouraging peer-to-peer support through recreational activities. The programme provides a safe space for students to talk about how chronic armed violence affects them, and how best to cope with it. The programme is taught to teachers so they can replicate it. A similar programme from Rio, the ICRC's Creating Humanitarian Spaces programme, has been replicated in Honduras and elsewhere (ICRC 2011).

In all these settings, the ICRC works closely with weapons bearers from state and non-state actors. The ICRC's singular humanitarian purpose and its principled commitment to impartiality, neutrality, and independence, combined with long experience of its own risk management in volatile environments, means the ICRC is able to engage directly with armed groups of various kinds, and with the government security forces. This enables it to address the humanitarian needs of affected populations with these weapons bearers and strive to influence their behaviour towards local communities and essential urban services.

A humanitarian agenda for safer cities

The ICRC's experience in these two distinct types of violent setting – armed conflict and chronic urban violence – makes clear the vital importance of the continuity of safe urban services across both these very different situations. In armed conflict and chronic urban violence, people's safe access to essential services is critical to their ability to survive and thrive. The continuity of these services is also fundamental to IHL, IHRL, and any hope of achieving the goals of the 2030 Agenda for Sustainable Development.

This chapter argues that the humanitarian agenda for safe access to services in cities is clear. Protecting people requires respect for law and a combined approach to humanitarian operations that meets people's urgent needs as violence flares and spikes across different urban spaces, while simultaneously remaining focused on sustaining the resilience of essential urban infrastructure and services. These recommendations are born out of the ICRC's experience in urban areas, and must form the basis for cooperation and programme design between national authorities, municipal authorities, communities, and humanitarian organizations.

References

Auyero, J., P. Bourgois, and N. Scheper-Hughes (eds). 2015. *Violence at the Urban Margins*. New York: Oxford University Press.

Campbell, L. (2016) 'Stepping Back: Understanding Cities and their Systems', ALNAP Working Paper. London: Overseas Development Institute.

Cue, W. and V. Nunez-Flores. 2017. 'According to Need? Humanitarian Responses to Violence in Central America', June, available at http://odihpn.org/magazine/according-need-humanitarian-responses-violence-central-america (accessed 26 March 2018).

Geneva Declaration Secretariat. 2015. *Global Burden of Armed Violence: Every Body Counts*. Geneva: Geneva Declaration Secretariat.

George, A. R. 1999. *The Epic of Gilgamesh: The Babylonian Epic Poem and Other Texts in Akkadian and Sumerian – Translated and with an Introduction by Andrew George*. London: Allen Lane.

ICRC (International Committee of the Red Cross). 2011. 'Creating Humanitarian Spaces Enters Second Phase in Brazil's Schools.' Geneva: ICRC, available at www.icrc.org/eng/resources/documents/feature/2011/aeh-2011-04-12.htm (accessed 3 July 2018).

ICRC (International Committee of the Red Cross). 2013. *Professional Standards for Protection Work Carried Out by Humanitarian and Human Rights Actors in Armed Conflict and Other Situations of Violence*. Geneva: ICRC.

ICRC (International Committee of the Red Cross). 2015. *Urban Services During Protracted Armed Conflict: A Call for a Better Approach to Assisting Affected People*. Geneva: ICRC.

ICRC (International Committee of the Red Cross). 2016a. 'Protracted Conflict and Humanitarian Action: Some Recent ICRC Experiences', September. Geneva: ICRC, available at www.icrc.org/en/document/protracted-conflict-and-humanitarian-action (accessed 26 March 2018).

ICRC (International Committee of the Red Cross). 2016b. 'Rio Project: Programmes Help to Reduce the Humanitarian Impact of Urban Violence', November. Geneva: ICRC, available at www.icrc.org/en/document/rio-project-programs-helped-reduce-humanitarian-impact-violence-rio-de-janeiro (accessed 26 March 2018).

ICRC (International Committee of the Red Cross). 2016c. 'Armed Violence and the New Urban Agenda: The ICRC's Recommendations for HABITAT III', June. Geneva: ICRC, available at http://blogs.icrc.org/law-and-policy/wp-content/uploads/sites/102/2016/07/ICRC-Policy-Paper-for-Habitat-III-final_LR.pdf (accessed 26 March 2018).

ICRC (International Committee of the Red Cross). 2017. *'I Saw My City Die': Voices from the Front Lines of Urban Conflict in Iraq, Syria and Yemen*. Geneva: ICRC.

Koonings, K. and D. Krujit. 2015. *Violence and Resilience in Latin American States*. London: Zed Books.

Mark, J. 2014. 'The Ancient City', in *Ancient History Encyclopedia*, 5 April, available at www.ancient.eu/city (accessed 26 March 2018).

NRC and IFRC (Norwegian Refugee Council and International Federation of Red Cross and Red Crescent Societies). 2016. 'The Importance of Addressing Housing, Land and Property (HLP): Challenges in Humanitarian Response.' Oslo and Geneva: NRC and IFRC, available at: www.nrc.no/globalassets/pdf/reports/the-importance-of-housing-land-and-property-hlp-rights-in-humanitarian-response.pdf (accessed 26 March 2018).

Osman Ali, O. and A. Mahmoud. 2016. 'From a Temporary Emergency Shelter to an Urban Neighbourhood: The Abu Shoak IDP Camp in North Darfur', Sudan Working Paper No. 3. Bergen: Chr. Michelsen Institute, available at www.cmi.no/publications/file/5928-the-abu-shoak-idp-camp-in-north-darfur.pdf (accessed 26 March 2018).

UN-Habitat (UN Human Settlements Programme). 2016. *The New Urban Agenda*. Nairobi: UN-Habitat.

UNSC (United Nations Security Council). 2016. Resolution 2282, 'Post-Conflict Peacebuilding', UN Doc. S/RES/2282, 27 April. New York: United Nations.

WHO (World Health Organization). 2012. *Femicide: Understanding and Addressing Violence against Women*. Geneva: WHO, available at http://apps.who.int/iris/bitstream/10665/77421/1/WHO_RHR_12.38_eng.pdf (accessed 26 March 2018).

World Bank. 2011. *World Development Report: Conflict, Security and Development*. Washington, DC: World Bank.

World Bank. 2016. 'Urban Violence: A Challenge of Epidemic Proportion', 6 September. Washington, DC: World Bank, available at www.worldbank.org/en/news/feature/2016/09/06/urban-violence-a-challenge-of-epidemic-proportions (accessed 26 March 2018).

World Bank. 2017. 'Somalia: World Bank Approves $50 Million in Support of Drought Response and Recovery', 30 May. Washington, DC: World Bank, available at www.worldbank.org/en/news/press-release/2017/05/30/somalia-world-bank-approves-us50-million-in-support-of-emergency-drought-response-and-recovery (accessed 26 March 2018).

9 Daring the unconventional on the pathways for peace

On the 'how' of sustaining peace in the city

Achim Wennmann

Introduction

Children of Puerta Real start their day by jumping over a wall on their way to school. This wall separates their gated community – about 30 kilometres south of the centre of Guadalajara, Mexico's second-largest city – from the one next door, Cima del Sol. The physical barrier was a practical way for the residents of Cima del Sol to maintain their property value at an average purchase price of US$17,500 per unit, compared to the lower rate of US$15,000 in Puerta Real. It was also a practical solution to reduce traffic in Cima del Sol because Puerta Real residents kept transiting through their community, despite having a separate entrance. Beyond cutting off access to the public school, the result was also a social division between those slightly better off (with household incomes of around US$8 per day) and those somewhat worse off (Pfannenstein 2018).

Cima del Sol and Puerta Real are part of a broader trend for gated communities in Guadalajara. Pfannenstein *et al.* (2017) show that by 2015 15 per cent of the territory of the metropolitan area had been privatized, involving 233 luxury developments on larger plots and 2,361 gated communities on smaller plots, mainly for the middle classes but also including micro-developments and social housing. Developments for the middle classes are frequently in close proximity to luxury developments to gain in value by association and facilitate access to services, while gated communities for lower-income households are mainly located far away from the city centre. This spatial distribution of gated communities in turn contributes to traffic problems, gentrification, and social fragmentation (ibid.: 175–177).

Experiences of gated communities similar to those in Guadalajara are found across the world's cities (Blakely and Snyder 1997; Caldeira 1996). While high-end communities keep a wealthier elite apart from the rest of society, gated communities are no longer only targeted at the rich but also the upper and lower middle classes. Gated communities translate walls that delimit territory into walls between people. Such divisions are not unknown to peacebuilders. Cities like Beirut, Belfast, Nicosia, and Jerusalem, for instance, 'tend to freeze the conflict and remain partitioned regardless of a conflict settlement. Thus the divided city demonstrates the power of place to peacebuilding', with the population in these cities becoming 'socially as well as spatially segregated as temporary barriers have become permanent, and imagined walls have become real' (Björkdahl 2013: 207, 208). The realities of divided cities and the resulting day-to-day experiences of exclusion can cement inequalities and marginalization into the urban space and social fabric of cities, as recognized by a new 'spatial turn in peace research' (Björkdahl and Kappler 2017: 137).

These characteristics of cities are a reality check for international agendas that seek to promote peaceful and inclusive societies. Negotiated by diplomats and shaped by multilateral consensus and power politics, the Sustainable Development Goals (SDGs), the New Urban Agenda (NUA), and the Sustaining Peace Agenda were conceived with little concrete consideration about how to take such ambitions beyond declaratory commitments. With the agendas defined, international actors have moved on to discussing 'implementation' (United Nations and World Bank 2018; Steven 2017).

The analysis of this volume speaks to bridging the gap between international ambitions to foster peaceful and inclusive societies and actually achieving those ambitions in cities. While it is well understood that sustainable development will be won or lost in cities, it is much less clear how to win. Based on the Technical Working Group (TWG) on Urban Safety and Peacebuilding Practice described in the Introduction by Wennmann and Jütersonke, this volume aims to provide new perspectives on the conceptual and practical challenges and opportunities of sustaining peace in the city. It seeks to enrich the discussions on the implementation of the SDGs, the NUA, and the Sustaining Peace Agenda with conceptual reflections and practical experiences at the interface between urban safety and peacebuilding practice.

This chapter draws together some of the main themes of the book. By emphasizing that the confluence of urban safety and peacebuilding practice can result from the practical needs to solve problems in specific places, the chapter explores the options for achieving safer and more inclusive cities. It shows how both urban safety and peacebuilding professions can look back at a long record of driving participatory processes and working within contentious political economies. The degree of local agency for change indicates the existence of spaces that frame and give oxygen to transformative processes aiming to reduce violence or build peace in cities. These spaces exist in different shapes and forms, but there is much expertise on how to create and maintain them as a means to drive negotiated change. This operational focus brings into the fold a series of unconventional actors, processes, and ways of working that evolve from the alignment of partisan interests on the pathways for peace. By garnering know-how about working politically, urban safety and peacebuilding professionals have much to contribute to the coproduction of safer and inclusive cities.

On the confluence of urban safety and peacebuilding practice

The first theme of this volume focuses on the question of how to make sense of the *confluence* of urban safety and peacebuilding practice. In Chapter 1 Austin and Jütersonke describe how – from the perspective of minimalist definitions – peacebuilding practice has evolved from a focus on tackling the root causes of conflict to efforts seeking to transform conflict settings through direct engagement with the perpetrators of violence and the factors that shape their behaviour. In contrast, urban safety approaches target enhanced planning, governance, and spatial management to foster safer cities. Urban safety programmes thus zoom in on the material environment within which conflict and violence occurs. With reference to public health, Austin and Jütersonke argue that urban safety approaches primarily take on palliative roles:

> Palliative care *relieves* the symptoms of illness or, in this case, of urban crime, violence, and societal conflict. The goal of urban safety is similarly to *improve quality of life* for all those living in the city, without necessarily … working to cure the underlying problems that are causing these harms.

Peacebuilding, by comparison, takes the form of curative care that 'seeks to tackle the underlying causes of harm, whatever they may be'. Following such a minimalist view, urban safety practices focus on the technical exercise of reconfiguring urban spaces and reforming urban governance to minimize harm, nurture safety, and prevent crime; while peacebuilding approaches focus on the political work to change the behaviour of actors believed responsible for causing conflict.

While the minimalist, ideal-type definitions offered by Austin and Jütersonke tease out the essential *differences* between the two approaches, ' "field realities" in cities have … brought together both researchers and practitioners from the urban safety and peacebuilding communities by simple necessity'. Quoting the *White Paper on Peacebuilding* (Geneva Peacebuilding Platform 2015: 17), they note that 'at the field level, many people simply get on with doing what is needed to build peace and do not worry about definitions'. Instead, and to make sense of the confluence of the two fields of practice, it may be more insightful to understand how practice evolves from 'patterns of consent and disagreement' in a particular place rather than from the perspective of 'preconceived analytical categories' such as peacebuilding or urban safety (ibid.).

Allowing practical responses to constitute themselves in a specific context – or from the bottom up – offers new opportunity for innovative responses to the multiple harms, hazards, and threats that cities and their residents are facing. For instance, Salahub and Zaaroura in Chapter 6 in this volume

> … identify where and how taking gender as the starting point for analysis – rather than compartmentalizing it or adding it on as one of many layers of analysis – can strengthen the potential of the urban safety–peacebuilding nexus to prevent and reduce violence.… Gender analyses would allow the peacebuilding and urban safety practice community to overcome the strong focus on large-scale violence and turbulent political transitions that peacebuilding and many planning-based urban safety interventions tend to target

> and include the more mundane, less physically harmful violence that – arguably – more people experience on a daily basis in their households.

They also point to the importance of recognizing the heterogeneity in society between men and women, and boys and girls, at the city level; and highlight that assuming programmes and policies have the same effects on different gender groups misses practical opportunities for promoting safer and inclusive societies. Such opportunities may arise when adopting a view from the street level upward, where the primary transformative potential rests with women; hence correcting the cliché that men are perpetrators and women are victims. This transformative potential is reflected in the ever-growing claim of women to participate in decision-making, as evidenced in the trend towards participatory mechanisms surrounding peace negotiations. In this context, neighbourhood and local politics already appear to be domains in which women are taking on political roles more frequently, in comparison to national and international levels. Yet Salahub and Zaaroura also stress the barriers that stand in the way of allowing such transformative potential to come to full fruition, given the power of patriarchal attitudes that 'reinforce existing exclusive and exclusionary power structures'.

In Chapter 7 Homel and Masson provide another example of programming from the bottom-up whereby practice was adapted to a particular context and made to work despite broader systems that frustrate change. In their study on security sector reform (SSR) in the Governorate of Jenin of the Occupied Palestinian Territories, they show how a bottom-up, community-based approach builds effective partnerships for crime prevention with both formal and informal security providers to produce a workable mechanism for a safer community with stronger local leadership. They recall that 'since the 1980s, long before international aid agencies started promoting local community safety programmes in fragile contexts, criminologists had already questioned the efficiency of "hard" top-down responses to address citizens' insecurity'. In the case of Jenin, they argue that the

> … gap between the people's security needs and expectations on the one hand and the security agenda of the international community on the other generated mutual mistrust, undermined the reputation of the Palestinian Authority, eroded the credibility of its security forces, and remained the most fundamental obstacle to the SSR programme's sustainability. The community engagement process that was integral to the local crime prevention planning process worked to overcome this strategic deficit.

The participatory approach to designing partnerships for crime prevention across divided communities with a long history of abuse and mistrust provided 'a forum for galvanizing the development of at least a shared, if not always common, purpose'.

Overall, the chapters on gender and SSR illustrate the importance of participatory approaches and mechanisms to foster solutions from the bottom up by promoting local ownership – by men and women – and adapting and fusing practice developed in specific silos, such as crime prevention, SSR, urban safety, and peacebuilding. Participation and the bottom-up way of working are recurring themes in this volume, and lie at the heart of the conversation about *how* to go about achieving safer and more inclusive cities.

Situating the *how* in urban safety and peacebuilding practice

Writing from the perspective of architecture, in Chapter 2 Lopéz Rueda and Cara highlight the merits of reflecting upon 'encounter' and 'enclosure' models that shape inclusive or exclusive urban configurations. Building on the seminal work by Jane Jacobs (1961), they highlight how encounter approaches to urban design create open and permeable structures that favour interactions between residents of a neighbourhood and visitors from elsewhere. Designing urban configurations that entail short blocks with multiple connections enables spaces that nurture relationships between people, and where inclusive societies can thus grow. In contrast, Lopéz Rueda and Cara note that the enclosure model 'is an explicit measure of social control that evolved naturally towards the gated community'. It creates a sense of territoriality through walls and defences, and therefore creates separation between residents and those from outside (see also Dovey 2000).

Both encounter and enclosure models stand in the tradition of urban designers, planners, and architects focusing on the reconfiguration of the built-up environment to foster more – or less – inclusive social relations. However, Lopéz Rueda and Cara point out that the contemporary urban studies literature tends to refute deterministic approaches which hold that 'the environment can determine or cause certain kinds of behaviour'; hence the assumption that 'if the built environment is changed, behaviour will also be improved' is not immediately vindicated. If urban designers, planners, and architects want to foster safer cities, they cannot do so by themselves. They need partnerships with other communities of practice that are more directly specialized in targeting and transforming the behaviour of people. This specialization can be found in the field of peacebuilding.

The broader peacebuilding community represents a range of actors using 'dialogue, trust-building, and consensus-seeking processes to resolve or manage conflict through non-violent means' (Geneva Peacebuilding Platform 2015: 6). Over the last decade the peace mediation and conflict resolution profession has evolved into a discreet global web of experts with channels of communication to any armed actor – whether state security forces, rebel groups, or other perpetrators of violence. These networks do the important exploratory work to see if parties are ready for talks or will allow access for humanitarian assistance into the territory they control. They can provide expertise to and accompany parties during a peace process, including helping to navigate the legal grey space when communication or negotiation with 'terrorists', 'criminals', or 'gangs' is criminalized as part of war-fighting, anti-terror, or anti-drug policy. Over the years mediation support has become mainstream – there is even *United Nations Guidance on Effective Mediation* (United Nations 2012) – and is supported by many international organizations, governments, and private foundations.

Yet mainstream peace mediation seems trapped in understanding conflict in terms of interstate and civil wars, and thus from a state-centric perspective. City-level engagements can then only be perceived under the glass floor of national processes. Countering this tendency means building new networks and relationships, and envisioning and standing up for the use of the 'political' instruments of peace mediation and conflict resolution as a primary response to what is frequently described as 'criminal violence' (Wennmann 2018). At present the policy space for transforming the behaviour of people – especially those challenging existing orders or advancing interests through violence – is still largely influenced by a law-and-order approach. However, excessive reliance on this approach has in many cases generated very limited results in terms of violence reduction and inclusive development (Jütersonke *et al.* 2009).

Such dynamics may explain why 'the design of public space is left in a strange position of having police officers, rather than architects and urban designers, be responsible for spatial configuration', as noted by Lopéz Rueda and Cara. While this observation underlines the *who* in nurturing inclusive urban societies and the distribution of roles and responsibilities, it also raises the *how*, questioning the mantra of partnership to achieve collective goals in places where power relations among key actors are asymmetric. Such asymmetry should come as no surprise, given the many overlapping systems of governance in most cities beyond the trappings of the officially recognized authorities. These constellations – also called 'hybrid political orders' – recognize the fact that territorial control and city spaces are shaped and governed by a multitude of actors (Boege *et al.* 2009).

> While illicit or illegal activities may be illegitimate in the eyes of governing organisations, local communities and individuals may view them in entirely different ways. In the absence of governing institutions that meet their needs, local populations may view criminal enterprises as having legitimacy and inspiring greater loyalty.
>
> (Shaw and Howell 2016: 13)

Hence strategies that focus exclusively on the official local government or governance may fail to understand the entry points for change at the city level. These entry points are more frequently found with the actors holding *de facto* power.

Working within contentious urban political economies

Each city has its own political constellations that evolve in light of stresses, crises, or the agency behind specific interests. In this sense, each city has its own 'political settlement', defined as the arrangements between elites and the related formal or informal institutions that manage violence, distribute wealth and power, and create order. Yet these constellations change over time, as power holders continuously negotiate a new *mix* of wealth, order, justice, and freedom to decide on the distribution of 'costs and benefits, risks and opportunities to social groups, enterprises and organizations within the system' (Strange 1988: 24). Research for the United Kingdom's Stabilisation Unit places renewed emphasis on the centrality of elite bargains and political deals for managing violence and shaping post-war transitions. The research holds that 'an understanding of underlying power relations is an essential starting point for effective policy and planning' (Cheng *et al.* 2018: 1). The third theme of this volume speaks to these dynamics and the ability to advance safer and inclusive cities within contentious urban political economies.

The realities of unequal power relationships and control over economic resources bring a set of new actors into the discussion about sustaining peace in the city. From this negotiation perspective, actors such as gangs, police forces, disenfranchised populations, property developers, or other de facto power holders – any party with a stake in the outcome – must somehow be included in the process of building consensus around a shared vision for safety, security, and peace (Wennmann 2011). Van den Eertwegh in Chapter 4 in this volume demystifies such interactions and the use of dialogue and negotiation with criminal armed groups in urban settings. He highlights that such practices are more common than one might assume if considering, for instance, the use of plea bargaining, the engagement of informers, kidnap negotiations, and gang truces. This practice stands to contradict the narrative that authorities do not negotiate with 'terrorists' or 'criminals'. A more explicit use of dialogue and negotiation mechanisms expands the toolbox for mayors and community leaders to reduce violence and build an inclusive peace (see also Wennmann 2018).

An increasing amount of research focuses on the use of dialogue and negotiation to exit criminal violence (Wennmann 2014; Cockayne *et al.* 2017). Case experience includes, for instance, the Los Angeles gang truce (Umemeto 2006), the gang truce in El Salvador (Tager and Umaña 2013; von der Borgh and Saventije 2014), and the social inclusion of gangs in Ecuador (Brotherton and Gude 2018). As Austin and Jutersonke highlight in Chapter 1, increasing numbers of initiatives draw inspiration from the insight of public health experts, who have noted that violence spreads like a disease and is therefore amenable to strategic interruption points. Programmes deploy trusted members of a local community – from ex-drug runners or gang leaders to religious figures or elders – as 'violence interrupters'. These community members are trained to intervene in crises, mediate disputes among individuals, and intercede in group disputes to prevent violent events (Slutkin *et al.* 2014). The role of interrupters in detecting and diffusing conflicts, identifying and treating the highest-risk individuals, and changing social norms is well documented in evaluations showing a measurable reduction in violence due to such interrupter programmes (Skogan *et al.* 2009).

The experience of peace mediation similarly underlines the value of 'insider' mediators. These mediators are often leaders of civil society organizations, churches, trade unions, or business councils who leverage trust, respect, and deep knowledge of the dynamics and context of the conflict in conjunction with a high level of legitimacy rooted in their social position, personality, and skills. The fact that local leaders acting as mediators are connected to, and trusted by, important local constituencies has been found to build trust in processes and outcomes where the state is too weak or illegitimate to do so, de-escalating and managing conflict risks (Dziatkowiec 2017; Mason 2009).

What emerges from many evidence-informed interventions against urban violence and crime is that dialogue and negotiations should be part of a broader, integrated strategy. For instance, the application of focused deterrence strategies in the United States has generated 'noteworthy crime reduction impacts and should be part of a broader portfolio of crime reduction strategies available to policy makers and practitioners' (Braga *et al.* 2018: 1). Such strategies aim to change offender behaviour by understanding the dynamics and conditions that produce crime and sustain recurring crime problems, and by implementing a focused and integrated strategy of law enforcement, community mobilization, and social service actions (Kennedy 2008). This call for integrated strategies is mirrored by assessments of violence reduction strategies highlighting that promising approaches are those that 'bring together a range of violence prevention and reduction strategies across a number of sectors and purposefully target the key risk factors' of conflict and violence, and 'integrate [armed violence reduction and prevention] objectives and actions into regional, national, and subnational development plans and programmes' (Eavis 2011: 57–58; see also UNDP 2013; Comunidad Segura 2011; OECD 2011; Jaitman and Guerrero Compeán 2015). At the multilateral level, the diversity of this case experience is becoming progressively recognized as a strategy against violence in highly insecure areas (OAS 2015).

The call for integrated strategies is also emphasized in relation to countering extortion – which, according to Reitano and Shaw in Chapter 5, is one of the most understudied yet most pervasive phenomena of urban security. Extortion is the number one concern of many mayors, as it can affect all aspects of life and shape cultures of insecurity and fear. To expand policy options, Reitano and Shaw recognize extortion as a socioeconomic and political phenomenon, rather than exclusively as a security or justice challenge. Extortion is 'a strategy of local criminal governance' and countering it requires 'strong mobilization and solidarity from key business and community figures' and efforts 'to change the social norms around extortion'. Crucially, responses may need to be more hybrid and 'integrate, legalize, and regulate the criminal market itself, thereby driving out the fluid grey space that extortion groups move into'. In urban political economies it is therefore important to focus on core partisan interests rather than labels of the parties, and to accept that exits from extortion are built around outreach to atypical actors and atypical strategies (Wennmann and Ganson 2016). Yet the UK Stabilisation Unit's study on elite bargains and political deals underlines that it is exactly this type of understanding which 'is frequently lacking amongst external intervenors', and that 'interventions can be ineffectual, or counter-productive, when intervenors fail to analyse and engage effectively with underlying configurations of power and processes of elite bargaining' in conflict-affected states (Cheng *et al.* 2018: 1).

Coproducing safer cities

When looking at the work of urban designers, planners, and architects (which seeks primarily to shape the configuration of cities) in contrast to the work of peacebuilders (with a main focus on changing behaviour), it becomes clear how both sets of practice still advance their strategies largely within the confines of their respective professions. Yet neither community will be able to foster safer and more inclusive cities if they stay within their silos. This is why it is important to discern practical avenues to foster collaborative mechanisms to coproduce safe and inclusive cities.

Entry points for such collaboration across urban safety and peacebuilding communities can take many forms. As highlighted by Austin and Jütersonke in Chapter 1, such entry points tend to emerge from the practical realities faced by 'doers' in specific settings, where practical responses emerge through processes of adaptation, and trial and error. Initial work on the topic of entry points highlights opportunities across a range of topics and practices (Grob *et al.* 2016). These include a focus on topics where urban safety and peacebuilding professionals have complementary know-how. A range of technical issues – such as process design, SSR, constitution making, transitional justice, and monitoring and accountability – can enable exchanges at the technical level between urban safety and peacebuilding professionals which can then be adapted to city-level programming and decision-making in specific settings.

An underling enabler for coproducing safety is the use of participatory practices in both urban safety and peacebuilding communities. In fact, both communities can look back at decades of experience in participatory process design. Milliken (2014: 6) observes that 'participatory tools have come to urban safety from the development community, where methods to gather information by, about and with local people have been in use since the 1970s'. Among urbanists, such practices have been applied to urban planning processes. This participatory planning entails a 'systematic effort to envision a community's desired future, plan for that future and involve and harness the specific competencies and inputs of community residents, leaders, and stakeholders in the process' (Beyae 2009: 58). The aim is to build consensus around common goals and a vision for a neighbourhood that offer an alternative to hierarchical planning mythologies (McTague and Jakubowski 2013: 183–184).

Practices of this sort resonate with the broader peacebuilding community in various ways. In the conflict resolution literature such processes are part of the 'forward-looking' dimensions of peacemaking that capture the visions for the future and paths towards new political, economic, or societal orders (in contrast to 'backward-looking' peacemaking that deals with past violence and injustices) (Zartman 2005: 295). Such participatory processes can also be part of vision-based politics and negotiations across a range of stakeholders to nurture buy-in (Wennmann 2010: 221). These efforts may also entail national dialogues that are 'negotiating mechanisms intended to expand participation in political transitions beyond the political and military elites' (Papagianni 2014: 1). Recognizing that transitional mechanisms must be put in place to compensate for those of the formal government that lack sufficient legitimacy, they foster 'a shared understanding among key political actors on principles' (ibid.: 11). Participation also connects to decades of community or grassroots work of peacebuilders, including localized initiatives often grouped under the label of 'civil society' peacebuilding (see van Tongeren *et al.* 2005). The primary achievement of these activities has been their 'role in increasing human security in settings where the dignity, safety, and voice of ordinary citizens are severely compromised or threatened' (Cortright *et al.* 2016: 231).

While at the international level the space for participation becomes increasingly curtailed (V-Dem Institute 2018), at the city level such participation may grow out of the practical need for urban authorities to reduce violence in the context of resource constraints and the pressures of urbanization. A seminal study on community-based urban violence prevention programmes highlights that public authorities become

> ... more receptive to considering joining forces even with the poorer sectors of the population and assigning duties and rights that formerly were the exclusive responsibility of the state, down to the community. Concrete examples include, among other activities, neighbours going on joint patrols with police officers or the police taking part in education prevention exercises with youth.
>
> (Mathéy and Matuk 2015: 7)

This research also points to a 'tendency worldwide to seek closer contact between civil society and the state in an effort to reduce urban violence' (ibid.).

This tendency is supported by the constantly proliferating knowledge and learning generated from local practices presented by Mathéy and Matuk (ibid.), as well as by the Global Network on Safer Cities (2014) in *100 Promising Practices on Safer Cities*. Concrete examples include projects in Cape Town on violence prevention through urban upgrading (Cooke 2017), the rehabilitation of San Salvador's central park Parque Cuscatlán (Glasswings 2016), and the 'Licence to Operate' approach to prevent and reduce gang violence in Los Angeles (Basheer and Hoag 2014). There is also a significant track record of ensuring inclusive and participatory cities through social space planning, as evidenced by the city of Cologne (Stadt Köln 2016). Chapter 8 in this volume by Hugo Slim, Caroline Putman-Cramer, Evaristo De Pinho Oliveira, and Rachel Banfield emphasizes local leadership in war-affected cities and the importance of local authorities in maintaining essential urban services. Given the interconnected and subterranean nature of much urban infrastructure, many local authorities and neighbourhood groups have maintained working relations across conflict lines to ensure service continuity and thereby mitigate the humanitarian impact of war.

These examples across different types of violent conflict demonstrate the degree of local agency. People living in difficult places do not just passively wait for violence to stop or peace to arrive, but go about shaping the life of their community on a daily basis.

Understanding spaces for sustaining peace

The degree of local agency points to the existence of spaces that frame and enable the work of reducing violence or building peace. Such spaces frequently stand in between approaches that focus on the urban environment through spatial reconfiguration or architectural models (described by the notions of 'safe house' or 'safe space' as physical environments for trusted exchanges) and approaches that focus on affecting the behaviour of actors through negotiation and dialogue (described by the notions of 'political space' or the idea of 'political oxygen' that enables discreet work across the lines of conflict). From an urbanist's perspective, the emphasis is on designing or delimiting space through physical structures; from a peacebuilder's perspective, the focus is on building relationships with actors to expand the political space for discreet dialogue in an attempt to alter behaviour.

Dennis Rodgers in Chapter 3 of this volume highlights that such spaces may come in unexpected shapes and forms, as he describes his own experience of an encounter of rival gangs in a market in the middle of Managua's gangland. This market area, as it turned out, was 'nobody's territory' and therefore not part of the turf war. As a gang member explained:

> ... it depends where you meet [the other gang members]. If they came into the *barrio*, then of course we'd beat the shit out of them and throw them out, it's our heartland, and if we'd met them in *colonia* Las Condes that would have been different too, as that's part of our operating territory, and you can't allow rival gang members to be there unchallenged if you come across them there. But the market is nobody's territory, so it's not the same.

This episode illustrates that even in some of the most violent places there can be neutral spaces that enable interaction and relationship building. Rodgers describes these as

> ... a space of exception within the broader gangland context, where different violent actors can coexist and engage with each other in ways that do not have to entail antagonism. This is important, because it clearly opens possibilities for concrete peacebuilding initiatives within gangland areas.

Experiences such as this one in Managua are not necessarily unique, but are frequently difficult for the outsider to discern. In Catuche – a violent neighbourhood in Caracas, Venezuela – a ceasefire between gangs emerged from within a local community centre that brought together a coalition of mothers who engaged the rival gangs in a truce in 2007. The truce eventually led to an agreement to coexist peacefully that has held so far (Fuente 2018). Another example is the Municipal Mediation Centre of Corregidora in Quérétaro, Mexico. This centre opens a space to mitigate local disputes before they escalate into violence or continue to fester. The centre involves local residents in the resolution of their problems and the monitoring of collaborative approaches to resolve them. It is a space to diffuse mediation and dispute resolution practice within the community, thereby fostering a culture of peace and respect (Municipio de Corregidora 2015: 36–40). Salahub and Zaaroura in Chapter 6 also recall the importance of the household as a space to reduce violence – mostly perpetrated by men against women – and as a fundamental unit from which to start building peaceful and inclusive societies. What these examples illustrate is the importance of constructing (and shaping the relationships within) spaces that can serve to nurture the coalitions necessary to achieve safe and inclusive cities.

Such spaces are described as 'city labs' in the discussions of the TWG. They are 'spaces for locally-led policy innovation to prevent and reduce violence and crime and to help build lasting relationships between people and authorities in urban settings' (UN-Habitat and Geneva Peacebuilding Platform 2014: 12). The rationale for city labs builds on the realization that many mayors 'understood that additional investment in coercive capability does not necessarily increase citizen security in the city' but also that

> … they frequently do not have access to the best practice and professional networks that combine urban safety and peacebuilding expertise. City Labs respond to this challenge by assisting locally-led policy innovation to prevent and reduce violence and crime and to help build lasting relationships between people and authorities, and by facilitating the exchange of best practice and learning across urban safety and peacebuilding expertise.
> (Ibid.: 1)

The underlying logic of city labs is to offer the necessary space to broker problem solving and coalition building across various stakeholders relevant to a specific setting. A starting point is to develop a precise analysis of the needs of citizens, as well as an understanding of social and conflict dynamics at the neighbourhood or street level. This analysis in turn provides the foundation for coalition building around specific activities to solve problems. The results of such priority identification may, however, sit uncomfortably with internationally defined priorities. Reitano and Shaw in Chapter 5, for instance, highlight that for many cities extortion is a much more pervasive challenge than terrorism or political violence. City labs can thus also be a strategy to build on local actors' superior understanding of local dynamics and protect against international advice or assistance that is all too often disconnected from local social and political realities (Anderson *et al.* 2013; Donini *et al.* 2005).

Yet efforts at problem solving and coalition building require professional and institutionalized support to coordinate and sustain them. Ad hoc processes convened directly by stakeholders can be stifled by the exhaustion of planning and managing complex collaborative initiatives that are outside the core mandate or expertise of any one participant. The G20 High-Level Panel on Infrastructure underlines that partnerships bringing diverse actors together 'require their own infrastructure' (see also World Bank 2014: 3). Ad hoc processes may also fall prey to wrangling among the players if one or another is perceived to be manipulating the process to achieve its preferred outcome. In social innovation research, such limits are well understood. Collective impact approaches, for example, highlight the importance of a 'backbone support organization' that provides services such as neutral facilitation or mediation, technology and communication, data collection and reporting, and administrative support as a critical enabler of complex collaborative efforts (Kania and Kramer 2011).

Conclusion

As discussion at the international level increasingly focuses on the implementation of the SDGs, the NUA, and the Sustaining Peace Agenda, it is timely to question *how* such implementation is really supposed to work. There is much faith in the enlightened leader in government, the municipality, business, civil society, and international organizations to advance *solution- and leader-driven change*. This approach usually entails 'a disciplined, formal project process' in which 'solutions are identified up-front and are the focus of change; the reform is fully planned; a champion drives the process; and a pure-form best practice solution is produced' (Andrews 2015: 197). The politics surrounding such a type of change process can be driven by a range of partisan interests, such as institutional interests to position an organization as the primary champion for implementation (to control resources and activities), or individual interests associated with personal career advancements or ego gratification.

The analysis in this volume, however, seems to paint a more nuanced picture of how change can happen in even the most difficult places. The chapters suggest a much more subtle process of *problem-driven iterative adaptation* that describes 'a process of experimentation and trial and error; with multiple agents playing different leadership roles; producing a mixed form of hybrid that is fitted a peculiar context' (ibid.). Such processes are cognizant of the fact that political change involves a high degree of uncertainty and fluidity. When reviewing the experience of transition from authoritarian rule in Latin America, we are reminded that a theory underlying such processes

> … would have to be a chapter in a much larger inquiry into the problem of 'underdetermined' social change, of large-scale transformations which occur when there are insufficient structural or behavioral parameters to guide or predict the outcome. Such a theory would have to include elements of accident and unpredictability, of crucial decisions taken in a hurry with very inadequate information, of actors facing irresolvable ethical dilemmas and ideological confusion, of dramatic turning points reached and passed

> without an understanding of their future significance. In other words, it would have to be a theory of 'abnormality', in which the unexpected and the possible are as important as the usual and the probable.
>
> (O'Donnell and Schmitter 1986: 3–4)

Three decades on from O'Donnell and Schmitter's observation, it may be uplifting to note that there is now a wealth of tested approaches to advance safe and inclusive cities in contexts of fluidity and contention (see Wennmann 2015). Many organizations working to diffuse turbulent transitions have matured, and operate in the practical realities with which they are confronted. Their work succeeds despite a complex sociopolitical environment, be that defined by social divisions, legacies of grievance, weak institutions, lack of trust in government, pressing socioeconomic challenges, or the presence of spoilers content to exploit or tolerate conflict to meet their narrowly defined interests (Wennmann and Ganson 2016). There is opportunity to apply this record of practice more systematically at the city level, but it means shifting the narrative from *diplomacy* (with the focus on achieving a unified high-level consensus) to *getting things done* (with a focus on resolving concrete problems in a specific place). Such an *operational* focus brings into the fold a series of unconventional actors, processes, and ways of working that often resort to finding solutions to problems through dialogue and negotiations. They focus on the engagement of the parties in conflict on their partisan interests, which in turn can foster unconventional alliances for change.

The practice and approaches discussed in this chapter offer a new promise for the orderly and peaceful management of problems, disputes, and violence in urban settings. They build on the principle of coproduction and partnership, but also on the importance of aligning partisan interests and incentives behind a vision for peaceful and inclusive cities. By combining expertise and know-how about shaping spatial configurations, creating political spaces, and driving participatory processes, urban safety and peacebuilding practice has much to offer to help sustaining peace in the city.

References

Anderson, M. B., D. Brown, and I. Jean. 2013. *Time to Listen: Hearing People on the Receiving End of International Aid.* Cambridge, MA: CDA Collaborative Learning Projects.

Andrews, M., 2015. 'Explaining Positive Deviance in Public Sector Reforms in Development', *World Development* 74, 197–208.

Basheer, A. and C. Hoag. 2014. *Peace in the Hood: Working with Gang Members to End Violence*. Nashville, TN: Turner Publishing.

Beyae, W. 2009. 'Place Making through Participatory Planning', in M. Forth (ed.) *Handbook of Research on Urban Informatics: The Practice and Promise of the Real-Time City*. Hershey, PA: IGI Global, 55–67.

Blakely, E. and M. Snyder. 1997. *Fortress America: Gated Communities in the United States*. Washington, DC: Brookings Institution Press.

Björkdahl, A. 2013. 'Urban Peacebuilding', *Peacebuilding* 1(2), 207–221.
Björgdahl, A. and S. Kappler. 2017. *Peacebuilding and Spatial Transformation: Peace, Space and Place*. London: Routledge.
Boege, V., A. Brown, K. Clemens, and A. Nolan. 2009. 'On Hybrid Political Orders and Emerging States: What is Failing – States in the Global South or Research and Politics in the West?', in M. Fischer and B. Schmelzle (eds) *Building Peace in the Absence of States: Challenging the Discourse on State Failure*. Berlin: Berghof Research Centre for Constructive Conflict Management, 15–31.
Braga, A. A., D. L. Weisburd, and B. Turchan. 2018. 'Focused Deterrence Strategies and Crime Control: An Updated Systematic Review and Meta-Analysis of the Empirical Evidence', *Criminology and Public Policy* 17(1), 1–46.
Brotherton, D. C. and R. Gude. 2018. *Social Inclusion from Below: The Perspectives of Street Gangs and Their Possible Effects on Declining Homicide Rates in Ecuador*. Washington, DC: Inter-American Development Bank.
Caldeira, T. P. R. 1996. 'Fortified Enclaves: The New Urban Segregation', *Public Culture* 8, 303–328.
Cheng, C., J. Goodhand, and P. Meehan. 2018. 'Securing and Sustaining Elite Bargains that Reduce Violent Conflict', synthesis paper. London: Stabilisation Unit.
Cockayne, J., J. de Boer, and L. Bosetti. 2017. 'Going Straight: Criminal Spoilers, Gang Truces, and Negotiated Transitions to Lawful Order', Crime–Conflict Nexus Series No. 5. Tokyo: United Nations University.
Comunidad Segura. 2011. *Good Practices in Armed Violence Reduction and Prevention: The Experiences of Civil Society in Five Regional Settings*. Rio de Janeiro: Comunidad Segura.
Cooke, J. (ed.). 2017. *Violence Prevention through Urban Upgrading: A Manual for Safety as a Public Good.* Cape Town: Violence Prevention through Urban Upgrading.
Cortright, D., M. Greenberg, L. Stone, and A. Milovanovic. 2016. 'Conclusion: Civil Society Empowerment for Human Security', in D. Cortright, M. Greenberg, and L. Stone (eds) *Civil Society, Peace and Power*. Lanham, MD: Rowman & Littlefield, 231–248.
Donini, A., L. Minear, I. Smillie, T. van Baarda, and A. C. Welch. 2005. *Mapping the Security Environment: Understanding the Perceptions of Local Communities, Peace Support Organizations and External Aid Agencies*. Medford, OR: Feinstein International Famine Center.
Dovey, K. 2000. 'Redistributing Danger: Enclosure and Encounter in Urban Design', *Australian Planner* 37(1), 10–13.
Dziatkowiec, P. 2017. *The Inside Story: The Impact of Insider Mediators on Modern Peacemaking*. Geneva: Centre for Humanitarian Dialogue.
Eavis, P. 2011. *Working against Violence: Promising Practices in Armed Violence Reduction and Prevention*. Geneva: Geneva Declaration Secretariat.
Fuente, A. 2018. 'Las Madres que Pusieron Paz en la Ciudad más Violenta del Mundo', *El Pais*, 17 January, available at https://elpais.com/elpais/2018/01/09/planeta_futuro/1515515628_702836.html (accessed 22 June 2018).
Geneva Peacebuilding Platform. 2015. *White Paper on Peacebuilding*. Geneva: Geneva Peacebuilding Platform.
Glasswings. 2016. 'Espacio Public Para la Ciudad que Queremos', presentation at Mantenimiento de la Paz en la Ciudad: Nuevas Perspectivas sobre la Seguridad Urbana, Quito, 18 October, available at www.gpplatform.ch/sites/default/files/GLASSWING%20-%20PARQUE%20CUSCATL%C3%81N_0.pdf (accessed 22 June 2018).

Global Network on Safer Cities. 2014. *100 Promising Practices on Safer Cities: Collation of Urban Safety Practices*. Montreal, Paris, and Nairobi: International Centre for the Prevention of Crime/European Forum for Urban Security/UN-Habitat.

Grob, A., A. Papadovassilakis, and L. Ribeiro Fadon Vicente. 2016. 'Entry Points for Programming across Urban Safety and Peacebuilding', Paper 16. Geneva: Geneva Peacebuilding Platform.

Jacobs, J. 1961. *The Death and Life of Great American Cities*. New York: Random House.

Jaitman, L. and R. Guerrero Compeán. 2015. *Closing Knowledge Gaps: Toward Evidence-based Crime Prevention Policies in Latin America and the Caribbean*. Washington, DC: Inter-American Development Bank.

Jütersonke, O., R. Muggah, and D. Rodgers. 2009. 'Gangs, Urban Violence, and Security Interventions in Central America', *Security Dialogue* 40(4–5), 373–397.

Kania, J. and M. Kramer. 2011. 'Collective Impact', *Stanford Social Innovation Review*, Winter, 36–41.

Kennedy, D. M. 2008. *Deterrence and Crime Prevention: Reconsidering the Prospect of Sanction*. London: Routledge.

Mason, S. (2009). *Insider Mediators: Exploring their Key Role in Informal Peace Processes*. Berlin: Berghof Foundation for Peace Support.

Mathéy, K. and S. Matuk. 2015. 'Introduction', in K. Mathéy and S. Matuk (eds) *Community-Based Urban Violence Prevention: Innovative Approaches in Africa, Latin America, Asia, and the Arab Region*. Bielefeld: Transcript, 6–22.

McTague, C. and S. Jakubowski. 2013. 'Marching to the Beat of a Silent Drum: Wasted Consensus-Building and Failed Neighbourhood Participatory Planning', *Applied Geography* 44, 182–191.

Milliken, J. 2014. 'Building Peace in Urban Settings: Lessons from the Ground and Recommendations for Collaboration', Paper 11. Geneva: Geneva Peacebuilding Platform.

Municipio de Corregidora. 2015. *Programmas de Éxito: Trabajando por una Cultura de la Prevención de la Violencia y la Delincuencia*. Corregidora: Municipio de Corregidora.

OAS (Organization of American States). 2015. *Searching for Common Approaches to Deal with Unconventional Conflicts and Violence in the Americas*. Washington, DC: OAS.

O'Donnell, G. and P. C. Schmitter. 1986. *Transitions from Authoritarian Rule: Tentative Conclusions about Uncertain Democracies*. Baltimore, MD: Johns Hopkins University Press.

OECD (Organisation for Economic Co-operation and Development). 2011. *Armed Violence Reduction: Enabling Development*. Paris: OECD.

Papagianni, K. 2014. *National Dialogue Processes in Political Transitions*. Geneva and Brussels: Centre for Humanitarian Dialogue and European Peacebuilding Liaison Office.

Pfannenstein, B. 2018. 'Gated Communities as Challenge for Inclusive Cities', presentation at World Urban Forum 9, Kuala Lumpur, 13 February.

Pfannenstein, B., E. E. Anacleto Herrera, and S. Sevilla Villalobos. 2017. 'La Ciudad Cerrada y su Diversificación Como Reto del Área Metropolitana de Guadalajara, Mexico', *Revista de Geografía Norte Grande* 68, 163–184.

Shaw, M. and S. Howell. 2016. *Governing Safer Cities: Strategies for a Globalized World*. Vienna: UN Office on Drugs and Crime.

Skogan, W. G., S. M. Hartnett, N. Bump, and J. Dubois. 2009. 'Evaluation of Ceasefire Chicago', 19 March, available at www.skogan.org/files/Evaluation_of_CeaseFire-Chicago_Main_Report.03-2009.pdf (accessed 22 June 2018).

Slutkin, G., C. Ransford, B. Decker, and K. Volker. 2014. *Cure Violence: An Evidence-Based Method to Reduce Shootings and Killings*. Washington, DC: World Bank.

Stadt Köln. 2016. *Stadtkonferenz 10 Jahre Lebenswerte Veedel*. Cologne: Stadt Köln.

Steven, D. 2017. *The Roadmap for Peaceful, Just and Inclusive Societies: A Call to Action to Change Our World*. New York: Center for International Cooperation, New York University.

Strange, S. 1988. *States and Markets*. London: Pinter.

Tager, A. G. and I. Aguilar Umaña. 2013. *La Tregua entre Pandillas: Hacia un Proceso de Construcción de Paz Social*. Guatemala City: Interpeace.

Umemoto, K. 2006. *The Truce: Lessons from the L.A. Gang War*. Ithaca, NY: Cornell University Press.

UNDP (UN Development Programme). 2013. *Citizen Security with a Human Face: Evidence and Proposals for Latin America*. New York: UNDP.

UN-Habitat and Geneva Peacebuilding Platform. 2014. 'City Labs for Safer Cities: Towards an Integrated Approach to Urban Safety and Peacebuilding', consultation draft, 24 March. Nairobi and Geneva: UN-Habitat and Geneva Peacebuilding Platform.

United Nations. 2012. *United Nations Guidance on Effective Mediation*. New York: United Nations.

United Nations and World Bank. 2018. *Pathways for Peace: Inclusive Approaches to Preventing Violent Conflict*. New York NY and Washington DC: United Nations and World Bank.

van Tongeren, P., M. Brenk, M. Hellema, and J. Verhoeven (eds). 2005. *People Building Peace II: Successful Stories of Civil Society*. Boulder, CO: Lynne Rienner.

V-Dem Institute. 2018. *Democracy for All? V-Dem Annual Democracy Report 2018*. Gothenburg: University of Gothenburg.

von der Borgh, C. and W. Saventije, 2014. 'De-Securitising and Re-Securitising Gang Policies: The Funes Government and Gangs in El Salvador', *Journal of Latin American Studies* 47(1), 149–176.

Wennmann, A. 2010. 'Conclusion', in M. Berdal and A. Wennmann (eds) *Ending Wars, Consolidating Peace: Economic Perspectives*. London: Routledge, 219–225.

Wennmann, A. 2011. *The Political Economy of Peacemaking*. London: Routledge.

Wennmann, A. 2014. 'Negotiated Exits from Organized Crime: Building Peace in Conflict and Crime-Affected Contexts', *Negotiation Journal* 30(3), 255–273.

Wennmann, A. 2015. 'Building Peace: An Annotated Resource Pack', Paper 14. Geneva: Geneva Peacebuilding Platform.

Wennmann, A. 2018. 'Crossing the New Frontier: Peace Mediation in the City', Oslo Forum Paper. Geneva: Centre for Humanitarian Dialogue.

Wennmann, A. and B. Ganson. 2016. 'Responses to Conflict that Work', Paper 15. Geneva: Geneva Peacebuilding Platform.

World Bank. 2014. *Overcoming Constraints to the Financing of Infrastructure: Success Stories and Lessons Learned*. Washington, DC: World Bank.

Zartman, I. W. 2005. 'Looking Forward and Looking Backward on Negotiation Theory', in I. W. Zartman and V. Kremenyuk (eds) *Peace Versus Justice: Negotiation Forward- and Backward-Looking Outcomes*. Lanham, MD: Rowman & Littlefield, 287–301.

Index